# Nanomaterials: An Introduction to Properties, Synthesis and Applications

# Nanomaterials: An Introduction to Properties, Synthesis and Applications

Emmanuel Craig

Larsen & Keller
www.larsen-keller.com

Nanomaterials: An Introduction to Properties, Synthesis and Applications
Emmanuel Craig
ISBN: 978-1-64172-106-6 (Hardback)

**⊟ Larsen & Keller**

Published by Larsen and Keller Education,
5 Penn Plaza,
19th Floor,
New York, NY 10001, USA

**Cataloging-in-Publication Data**

Nanomaterials : an introduction to properties, synthesis and applications / Emmanuel Craig.
        p. cm.
Includes bibliographical references and index.
ISBN 978-1-64172-106-6
1. Nanostructured materials. 2. Nanotechnology. I. Craig, Emmanuel.
TA418.9.N35 N36 2019
620.5--dc23

For more information regarding Larsen and Keller Education and its products, please visit the publisher's website www.larsen-keller.com

# Table of Contents

**Permissions**

**Index**

# Preface

Materials whose individual units range in size between 1 to 1000 nanometers are called nanomaterials. Such materials have unique mechanical, optical and electronic properties. These are either engineered or natural. Nanomaterials can be grouped under nanostructured materials and nano-objects. They have major applications in manufacturing processes, healthcare, bioimaging and biosensing, medical diagnosis, etc. The study of nanomaterials requires an integration of the fields of materials science and nanotechnology, advanced materials metrology and synthesis. This book is a compilation of chapters that discuss the most vital concepts in this field. Most of the topics introduced in this book cover the properties, synthesis, techniques and applications of nanomaterials. This textbook is a complete source of knowledge on the present status of this important field.

A short introduction to every chapter is written below to provide an overview of the content of the book:

**Chapter 1**, Any material whose size ranges between 1 to 100 nanometers is called a nanomaterial. It is studied under nanoscience and nanotechnology. The aim of this chapter is to provide an introduction to nanomaterials through the elucidation of the principles of nanoscience and nanaotechnology, and the sources of nanomaterials; **Chapter 2**, Nanoparticles are particles, which are between 1 and 100 nm in size that are surrounded by an interfacial layer. The characterization of the chemical and physical properties of nanoparticles is vital for nanotoxicology studies, exposure assessment and manufacturing process control. The topics elucidated in this chapter cover some of the important aspects of nanoparticles, such as the production and characterization of nanoparticles, magnetic and ferrite nanoparticles; **Chapter 3**, An understanding of nanomaterials requires an in-depth study of nanomaterial structure. The significant aspects of the structure of nanomaterials such as quantum dot, nanowire, nanorod, nanocrystalline materials, etc. have been extensively discussed in this chapter; **Chapter 4**, The objective of nanosynthesis is to produce a nanomaterial that exhibits desirable properties relative to the length scale of the materials. Accordingly, the synthesis should display control of size within the nano scale. Such methods of nanosynthesis are classified as bottom up and top down. The different aspects of nanoparticle synthesis, mechanical attrition, pyrolysis, etc. have been discussed in this chapter; **Chapter 5**, Nanomaterials are classified into groups, namely nanostructured materials and nano-objects. An elaborate study of the varied types of nanomaterials has been provided in this chapter, which includes topics such as carbon based nanomaterials, metal based nanomaterials, dendrimers and composites; **Chapter 6**, Nanometrology is a subfield of metrology, which involves the science of measurement at nanoscales. This chapter discusses the diverse measurement techniques such as atomic force microscopy, scanning tunneling microscopy, etc. for a holistic understanding of nanometrology; **Chapter 7**, Nanoparticles have potential applications in physics, electronics, optics and medicine. A lot of research is being conducted regarding the use of nanoparticles as potential drug delivery system and as dietary supplements for the delivery of biologically active substances. This is an important chapter, which will analyze in detail about such applications of nanoparticles in diverse industries.

I extend my sincere thanks to the publisher for considering me worthy of this task. Finally, I thank my family for being a source of support and help.

**Emmanuel Craig**

# Introduction to Nanomaterials

Any material whose size ranges between 1 to 100 nanometers is called a nanomaterial. It is studied under nanoscience and nanotechnology. The aim of this chapter is to provide an introduction to nanomaterials through the elucidation of the principles of nanoscience and nanaotechnology, and the sources of nanomaterials.

## Nanoscience and Nanotechnology

The term "nano" refers to the metric prefix 10-9. It means one billionth of something. "Nano" can be ascribed to any unit of measure. For example, you may report a very small mass in nanograms or the amount of liquid in one cell in terms of nanoliters.

So, what is nanoscience? Nanoscience is the study of structures and materials on the scale of nanometers. To give you an idea of how long a nanometer is, this printed page is about 75,000 nanometers thick. When structures are made small enough—in the nanometer size range—they can take on interesting and useful properties.

Nanoscale structures have existed in nature long before scientists began studying them in laboratories. A single strand of DNA, the building block of all living things, is about three nanometers wide. The scales on a morpho butterfly's wings contain nanostructures that change the way light waves interact with each other, giving the wings brilliant metallic blue and green hues. Peacock feathers and soap bubbles also get their iridescent coloration from light interacting with structures just tens of nanometers thick. Scientists have even created nanostructures in the laboratory that mimic some of nature's amazing nanostructures.

Because nanostructures are so small, specialized methods are needed to manufacture objects in this size range. Scientists use beams of electrons or ions to etch features as small as 25 nanometers into metal, silicon and carbon-based materials. In addition to being formed on these solid material surfaces, nanostructures can also be formed in liquids. Nanostructures can be created by reacting chemicals in liquids and gases to generate nanofibers, nanocrystals and quantum dots, some as small as one nanometer wide. Scientists are even learning how to build three-dimensional structures at the nanoscale. Called nano-electro-mechanical systems, or NEMS, these devices might one day be used like microscopic robots to carry out tasks too small for humans to do themselves. For example, NEMS could carry out surgery on a single cell or act as mechanical actuators to move around individual molecules.

In order to observe and study nanostructures, specialized equipment must be used. If you wanted to magnify something ten times, you could use a magnifying glass that fits in your pocket. If you wanted to magnify something 200 times, you would need a microscope that may weigh several pounds and take up part of a desk. To magnify nanoscale structures, high-powered microscopes that fill an entire room are needed.

## Nanoscience and Advanced Materials

The ability to create new technologies or devices would not be possible without the use of advanced materials. Energy is an important issue for any new device, and making devices smaller approaching the nano-scale can reduce the energy cost, while increasing speed. These nano-structures or nanodevices can behave in surprising ways which are not like miniaturized versions of the macroscopic devices. Ultimately this behavior is explicable by quantum mechanics, a branch of modern physics, but new methods of fabricating or interacting with such nano-structures is what nanoscience is all about, ideally to the benefit of technology and to people. Nanoscience incorporates applications in photonics, medical diagnostics, ultra-fast electronics and many other areas which in addition use advanced materials. Advanced materials include superconductors, polymers, lasers and optoelectronics and they can be found in applications ranging from computers and electronics, to telecommunications and broadcasting, to airlines and healthcare.

## Benefits of Studying the Behavior and Interactions of Nanoscale Materials

New behavior at the nanoscale is not necessarily predictable from that observed at large size scales. Important changes in behavior are caused not by the order of magnitude size reduction, but also by new phenomena such as size confinement, predominance of interfacial phenomena, quantum mechanics and Coulomb blockade. It is notable that all relevant phenomena at the nanoscale are caused by the tiny size of the organized structure as compared to molecular scale, and by the interactions at their predominant and complex interfaces.

## The Aims of Nanoscience

Once we are able to control feature size, we can enhance material properties and device functions beyond those that we currently know or even imagine. Nanotechnology aims to gain control of structures and devices at the atomic, molecular and supramolecular levels, and to learn how to efficiently manufacture and use these devices.

## Nanotechnology

Nanotechnology is a field of research and innovation concerned with building 'things' - generally, materials and devices - on the scale of atoms and molecules. A nanometer is one-billionth of a meter: ten times the diameter of a hydrogen atom. The diameter of a human hair is, on average, 80,000 nanometers. At such scales, the ordinary rules of physics and chemistry no longer apply. For instance, materials' characteristics, such as their color, strength, conductivity and reactivity, can differ substantially between the nanoscale and the macro. Carbon 'nanotubes' are 100 times stronger than steel but six times lighter.

Nanotechnology is hailed as having the potential to increase the efficiency of energy consumption, help clean the environment, and solve major health problems. It is said to be able to massively increase manufacturing production at significantly reduced costs. Products of nanotechnology will be smaller, cheaper, lighter yet more functional and require less energy and fewer raw materials to manufacture, claim nanotech advocates.

## Experts Views about Nanotechnology

In June 1999, Richard Smalley, Nobel laureate in chemistry, addressed the US House Committee on Science on the benefits of nanotechnology. "The impact of nanotechnology on the health, wealth, and lives of people," he said, "will be at least the equivalent of the combined influences of microelectronics, medical imaging, computer-aided engineering and man-made polymers developed in this century."

## Concerns about Possible Effects on Human and Environmental Health

Others, however, are as cautious as Smalley is enthusiastic. Eric Drexler, the scientist who coined the term nanotechnology, has warned of developing "extremely powerful, extremely dangerous technologies" Drexler envisioned that self-replicating molecules created by humans might escape our control. Although this theory has been widely discredited by researchers in the field, many concerns remain regarding the effects of nanotechnology on human and environmental health as well as the effect the new industry could have on the North-South divide. Activists worry that the science and development of nanotechnology will progress faster than policy-makers can devise appropriate regulatory measures. They say an informed debate must take place to determine the balance between risks and benefits.

## The Global Market for Nanotechnology Products

Given the promise of nanotechnology, the race is on to harness its potential - and to profit from it. Many governments believe nanotechnology will bring about a new era of productivity and wealth, and this is reflected by the way public investment in nanotechnology research and development has risen during the past decade. In 2002, Japan was dedicating US$750 million a year to the field, a six-fold increase on the 1997 figure.

## Estimates for the Value of the Nanotechnology Global Market

The US National Science Foundation predicts that the global market for nanotech-based products will exceed US$1 trillion within 15 years. Paul Miller, senior researcher at the British policy research organization Demos, said in 2002 that "already, roughly one-third of the research budgets of the biggest science-based firms in the US is going into nanotech" whilst the US National Nanotechnology Initiative's budget rose from US$116 million in 1997 to a requested US$849 million in 2004.

## Nanotechnology in the Developing World

In the developing world, Brazil, Chile, China, India, the Philippines, South Korea, South Africa and Thailand have shown their commitment to nanotechnology by establishing government-funded programmes and research institutes. Researchers at the University of Toronto Joint Centre for Bioethics have classified these countries as 'front-runners' (China, South Korea, and India) and 'middle ground' players (Thailand, Philippines, South Africa, Brazil, and Chile). In addition, Argentina and Mexico are 'up and comers': although they have research groups studying nanotechnology, their governments have not yet organized dedicated funding.

## Nanotechnology in Thailand and China

In May 2004, the Thai government announced plans to use nanotechnology in one per cent of all consumer products by 2013. Their market value by then is predicted to be 13 trillion baht (more than US$320 billion at contemporary exchange rates). Indeed, Thailand has wholeheartedly embraced nanotechnology and its development is a major commitment of the Thai government. Likewise, China announced in May 2004 that nanotechnology is central to its long-term national science and technology plan.

## Potential Benefits for Developing Countries

Nanotechnology holds the promise of new solutions to problems that hinder the development of poor countries, especially in relation to health and sanitation, food security, and the environment. In its 2005 report entitled Innovation: applying knowledge in development, the UN Millennium Project task force on science technology and innovation wrote that "nanotechnology is likely to be particularly important in the developing world, because it involves little labor, land or maintenance; it is highly productive and inexpensive; and it requires only modest amounts of materials and energy".

## Effects of Nanotechnology on Health and Sanitation

Nanotechnology is already useful as a tool in health care research. Some researchers used 'optical tweezers'- pairs of tiny glass beads are brought together or moved apart using laser beams - to study the elasticity of red blood cells that are infected with the malaria parasite. The technique is helping researchers to better understand how malaria spreads through the body.

## How Nanotechnology Might Improve Drug Delivery

But nanotechnology could also one day lead to cheaper, more reliable systems for drug-delivery. For example, materials that are built on the nanoscale can provide encapsulation systems that protect and secrete the enclosed drugs in a slow and controlled manner. This could be a valuable solution in countries that don't have adequate storage facilities and distribution networks, and for patients on complex drug regimens who cannot afford the time or money to travel long distances for a medical visit.

## Nanoscale Filters for Improved Water Purification Systems

Filters that are structured on the nanoscale offer the promise of better water purification systems that are cheap to manufacture, long-lasting and can be cleaned. Other similar technologies could absorb or neutralize toxic materials, such as arsenic, that poison the water table in many countries including India and Bangladesh.

## Food Security

## Using Nanosensors on Crops and Nanoparticles in Fertilisers

Tiny sensors offer the possibility of monitoring pathogens on crops and livestock as well as measuring crop productivity. In addition, nanoparticles could increase the efficiency of fertilizers. However, the Swiss insurance company Swiss Re warned in a report in 2004 that they could also increase the ability of potentially toxic substances, such as fertilizers, to penetrate deep layers of the soil and travel over greater distances.

## Using Nanotechnology Techniques to Grow Crops in Hostile Conditions

In addition, researchers in both developed and developing countries are developing crops that are able to grow under 'hostile' conditions, such as fields where the soil contains high levels of salt (sometimes due to climate change and rising sea levels) or low levels of water. They are doing this by manipulating the crops' genetic material, working on a nanotechnology scale with biological molecules.

## How Nanotechnology Methods Might Be used in the Areas of Renewable and Sustainable Energy to Help the Environment

The application of nanotechnology in the field of renewable and sustainable energy (such as solar and fuel cells) could provide cleaner and cheaper sources of energy. These would improve both human and environmental health.

## Nanoscale Filters and Nanoparticles could be used to Clean the Environment

Tiny wastewater filters, for example, could sift emissions from industrial plants, eliminating even the smallest residues before they are released into the environment. Similar filters could clean up emissions from industrial combustion plants. And nanoparticles could be used to clean up oil spills, separating the oil from sand, removing it from rocks and from the feathers of birds caught in a spill.

## Concerns about Nanoparticles and Toxicity

Research has shown that nano-sized particles accumulate in the nasal cavities, lungs and brains of rats, and that carbon nanomaterials known as 'buckyballs' induce brain damage in fish. Vyvyan Howard, a toxicologist at the University of Liverpool in the United Kingdom, has warned that the small size of nanoparticles could render them toxic, and warns that full hazard assessments are needed before manufacturing is licensed.

## Concerns about Nanoparticles in the Environment

Many interested parties, including the Canadian ETC Group and the insurance company Swiss Re, have expressed their concern over releasing tiny particles which, because of their small size, are able to travel very far into the environment. They warn that we do not yet know how these particles will act in the environment or what chemical reactions they will trigger on meeting other particles. However, these same groups also concur with nanotechnology advocates who feel the field may offer 'cleaner' technologies, and, ultimately, a cleaner environment. But mostly, the concern is for the lack of research into nanotechnology's potential threats to human health, society and the environment.

## Ensuring that Progress in Nanotechnology is Accompanied by Studies in Ethics and Societal Effects

In a paper published early in 2003, Anisa Mnyusiwalla, Abdallah Daar and Peter Singer, of the University of Toronto, Canada, wrote, "As the science of nanotechnology leaps ahead, the ethics lags behind. We believe that there is danger of derailing nanotechnology if serious study of its ethical, environmental, economic, legal and social implications does not reach the speed of progress in the science." According to Singer and his colleagues, in 2001, the US-based National Nanotechnology Initiative allocated US\$16-28 million to studying societal implications, but spent less than half that amount.

## Risk Assessment and Concerns Raised about Nanotechnology

Several non-governmental organizations are calling for greater risk evaluations or, in the case of Canada's ETC Group, a nanotech research moratorium. They, and others including the US-based Centre for Responsible Nanotechnology, have raised concerns about the following aspects of nanotechnology:

- The toxicity of bulk material, such as solid silver, does not help predict the toxicity of nanoparticles of that same material.

- Nanoparticles have the potential to remain and accumulate in the environment.

- They could accumulate in the food chain.

- They could have unforeseen impacts on human health.

- The public has not been sufficiently involved in debates on the applications, uses, and regulation of nanotechnology.

- Grey goo': Tiny robots generated with nanotechnology could acquire the ability to self-replicate.

- If the rich countries are the main drivers of the development of nanotechnology, applications which benefit developing nations will be side-lined.

- Unless rapid action is taken, research into nanotechnology could progress faster than systems can be put in place to regulate its applications and their uses.

## What the ETC Group Says about Nanotechnology

Although some of these concerns, mainly the 'grey goo' theory, have been widely discredited by researchers in the field, most remain high on the agenda of activists. The ETC Group has demanded that a UN moratorium be placed on all nanotechnology applications that could come into contact with the human body. The ETC Group has also expressed concern that the control of nanotechnology research and development might remain firmly in the hands of industrialized nations. The result would be a bias towards developing applications that benefit rich countries but neglect the needs of the poor.

## Looking at Nanotechnology from the Viewpoint of Developing Countries

"Significant nanotech activity is already occurring in developing countries," writes the UN Millennium Project task force on science technology and innovation in its 2005 report. "This activity may be derailed by a debate that fails to take account of the perspective of developing countries." The authors then caution that this activity could be ruined if public and policy debates fail to take account of the perspective of developing countries. At the time of writing, a global dialogue of stakeholders was underway to determine the potential impacts of nanotechnology on such countries.

## Nanoengineering - Engineering on the Molecular Scale

Advances in nanotechnology have built on advances in microscopy. As well as allowing molecules to be imaged, the Scanning Tunnelling Microscope allowed researchers to manipulate them by picking up and moving individual atoms. This is the essence of 'bottom up' or molecular nanotechnology - the notion that molecular structures can be built atom-by-atom.

## The Vision of Controlled Production at Molecular Level Via Self-replicating 'Assemblers'

Some claim that nanotechnology could ultimately lead to the miniaturization of controlled production to the molecular level in much the same way as happens in human cells when, for instance, enzymes break and rearrange bonds holding molecules together. The vision is of potentially self-replicating 'assemblers' - tiny devices operating in unison like miniature versions of factory assembling lines - to produce 'nanomaterials', new products that will revolutionize construction, medicine, space exploration and computing.

## The Nano-Conveyor Belt, 'DNA Robots' and Spinning Molecular Structures

The theory is well ahead of current realities and while some warn that self-replicating 'nanobots' pose an immense threat to humanity, others dismiss the idea as impossible. However, a recent production of a nano-conveyor belt that moves streams of particles rather than individual ones along a nanotube represents a major breakthrough, as does the development of a 'DNA robot' ten nanometers long capable of 'walking' along a pavement also made of DNA. Other significant developments are the discovery of spinning molecular structures, which herald the possibility of power generation and controllable motion at the molecular level.

## Top-down Production

In the 'top down' approach, which still dominates the field, pieces of material are machined and etched into nanoscale structures.

## Future of Nanotechnology

Nanotech knowledge is rapidly growing. The number of scientific publications in the field grew from about 200 in 1997 to more than 12,000 in 2002. Despite this, relatively few products using nanoparticles are currently on the market. On the whole, the ones that are already on sale do not address the issues highlighted above, of health, food security and the environment. Rather, they have focused on consumer applications that include improved sunscreens, crack-resistant paints and scratch-proof spectacle lenses. Like electricity and the internal combustion engine, nanotechnology is an enabling technology. As such, it is predicted to precipitate a range of innovations.

# Nanomaterials

Nanotechnologies involve designing and producing objects or structures at a very small scale, on the level of 100 nanometers (100 millionth of a millimeter) or less. Nanomaterials are one of the main products of nanotechnologies – as nano-scale particles, tubes, rods, or fibers. Nanoparticles are normally defined as being smaller than 100 nanometers in at least one dimension.

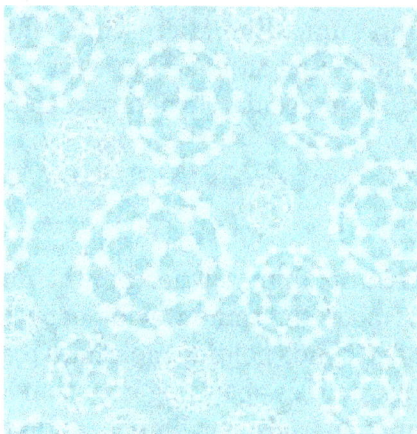

As nanotechnology develops, nanomaterials are finding uses in healthcare, electronics, cosmetics, textiles, information technology and environmental protection.

The properties of nanomaterials are not always well-characterized, and they call for risk assessment of possible exposures arising during their manufacture and use.

## Characterization of Nanomaterials

Descriptions of nanomaterials ought to include the average particle size, allowing for clumping and the size of the individual particles and a description of the particle number size distribution (range from the smallest to the largest particle present in the preparation).

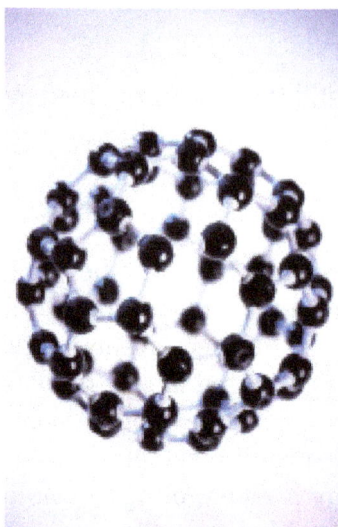

## Important Physical and Chemical Properties of Nanomaterials

The main parameters which are relevant to the safety of nanoparticles include:

### Physical Properties

- Their size, shape, specific surface area, and ratio of width and height.
- Whether they stick together.
- Size distribution.
- How smooth or bumpy their surface is.
- Structure, including crystal structure and any crystal defects.
- How well they dissolve.

### Chemical Properties

- Molecular structure.
- Composition, including purity, and known impurities or additives.
- Whether it is held in a solid, liquid or gas.
- Surface chemistry.
- Attraction to water molecules or oils and fats.

These parameters should be assessed both for the nanomaterials as they are made, and for nanomaterials as they are used (e.g. in the formulation in any subsequent product). Measurements need to take account of the way properties may change when nanomaterials are mixed with or suspended in other substances.

It is particularly important to measure how quickly nanomaterials dissolve in any liquid they are likely to come in contact with. Their small size means they dissolve faster than the comparable bulk (large) particulate material.

### Detection and Analysis of Nanomaterials

The best techniques for tracking nanoparticles depend on where they are. A variety of reliable instruments can identify airborne particles as small as a nanometer across. Electron microscopes can visualize nanoparticles in tissue slices and on the surface of materials. Nanoparticles in suspension in gas or liquids in some solid media can be measured by light scattering and by other microscopic techniques. The standard chemical analysis technique of mass spectroscopy - in which charged particles are separated by size - is applied to nanoparticles suspended in gases and liquids.

All these techniques have their strengths and weaknesses, and they may need to be used in combination. Most still need to be used in the laboratory by experienced technicians, but mobile and even hand-held equipment is becoming available especially for monitoring of air borne nanoparticles.

## Preparation of Nanomaterials for Biological Testing

Biological testing usually starts with a dry powder or a suspension in water or another liquid which is used to introduce nanoparticles into the organism. There are a range of suggestions for preparations which more closely resemble the way nanoparticles may be spread through a biological system. They include using the common proteins albumin, and fatty substances found in the lining of the lungs. Other possibilities include synthetic detergents, but these complicate assessments of toxicity.

## Measurement of Exposure to Nanomaterials

The measurement methods to use depend on the kind of exposure. The most reliable methods are for particles in the air. Nanoparticles may also be in contact with solids and liquids, especially in consumer products.

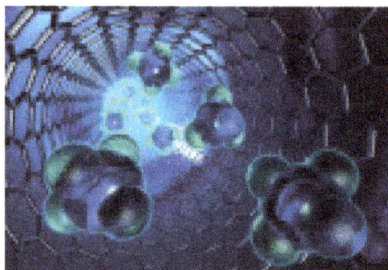

Current techniques to assess nanoparticle exposure are suitable for personal or area-based monitoring, continuous or discontinuous use, and basic characterization of samples. However, data on airborne exposures are scarce, and there have been few if any studies outside the workplace.

Exposure estimates from food and consumer products also remain difficult. Information on the presence of manufactured nanomaterials comes from manufacturers. There is also limited information about product use.

## Potential Health Effects of Nanomaterials

There is experimental evidence of a range of possible interactions with biological systems and health effects of manufactured nanoparticles. In experimental systems in the laboratory they can affect the formation of the fibrous protein tangles which are similar to those seen in some diseases, including brain diseases. Airborne particles might cause effects in the lungs but also on the heart and blood circulation similar to those already known for particulate air pollution. There is some evidence that nanoparticles might lead to genetic damage, either directly or by causing inflammation.

All these effects would depend on nanoparticles' fate in the body. Only a minimal amount of nanoparticle doses escape the lungs or intestine, but long-term exposure could still mean a large number are distributed round the body. Most are held in the liver or the spleen, but some appear to reach all tissues and organs. There may also be entry into the brain via the membranes inside the nose.

Nanotubes or rods with similar characteristics to asbestos fibers pose a risk of the mesothelioma (a form of cancer of the pleura).

## Potential Environmental Effects of Nanomaterials

Wider use of nanomaterials will lead to increases in environmental exposure. Little is known about how they may then behave in air, water or soil. They may be concentrated in particular "hot spots", either by clumping together with minerals or by interaction with organic matter.

Like other pollutants, they may pass from organism to organism, and perhaps move up food chains.

As a result of their diversity, nanomaterials may have a wide range of effects. Some kill bacteria or viruses. Experiments so far have also shown possible harmful effects on invertebrates and fish, including effects on behavior, reproduction and development. There is less research to date on soil systems and terrestrial species, and it is not clear whether laboratory results relate to what may happen out in the real world.

## How Well can we Assess the Risks from Nanomaterials

Existing risk assessment methods are generally applicable to nanomaterials but specific aspects related to nanomaterials need more development. They include methods for both estimating exposure and identifying hazards. The highest potential risks come from free, insoluble nanoparticles either dispersed in a liquid or as dust.

Risk assessment requires a detailed examination of properties, including:

- Particle size.

- Surface area.

- Stability.

- Surface properties.

- Solubility.

- Chemical reactivity.

Comparisons with well-known existing hazards may help inform risk assessment. They include those from airborne fine particles, and asbestos fibers.

The recommended approach to assess the risks from nanomaterials is still the four stage risk assessment proposed by the SCENIHR in 2007. Today, additional details can be added to this approach in the light of recent work on evaluating possible harmful effects of nanomaterials, especially using controlled laboratory tests (in vitro assays). These tests are useful for screening and for investigating mechanisms of adverse effects. However, tests using living organisms (in vivo assays) are also needed to improve knowledge of possible risks to people and the environment. Improvements are sought in the determination of exposures, and there is an urgent need for long-term exposure studies.

An OECD programme is producing dossiers on hazard identification for 14 common nanomaterials. Each will include physical and chemical properties, environmental effects, toxicology in mammals and material safety. This will help assess whether current OECD guidelines on identifying hazards are suitable for nanomaterials.

As knowledge improves, a category-based system to classify new nanomaterials may be developed, but at present a case by case approach is needed, leading to a data bank of case histories.

## Applications of Nanomaterials

## Nanomaterial Applications using Carbon Nanotubes

Applications being developed for carbon nanotubes include adding antibodies to nanotubes to form bacteria sensors, making a composite with nanotubes that bend when electric voltage is applied bend the wings of morphing aircraft, adding boron or gold to nanotubes to trap oil spills, include smaller transistors, coating nanotubes with silicon to make anodes the can increase the capacity of Li-ion batteries by up to 10 times.

## Nanomaterial Applications using Graphene

Applications being developed for graphene include using graphene sheets as electrodes in ultracapacitors which will have as much storage capacity as batteries but will be able to recharge in minutes, attaching strands of DNA to graphene to form sensors for rapid disease diagnostics, replacing indium in flat screen TVs and making high strength composite materials.

## Nanomaterial Applications using Nanocomposites

Applications being developed for nanocomposites include a nanotube-polymer nanocomposite to form a scaffold which speeds up replacement of broken bones, making a graphene-epoxy nanocomposite with very high strength-to-weight ratios, a nanocomposite made from cellulous and nanotubes used to make a flexible battery.

## Nanomaterial Applications using Nanofibers

Applications being developed for nanofibers include stimulating the production of cartilage in damaged joints, piezoelectric nanofibers that can be woven into clothing to produce electricty for cell phones or other devices, carbon nanofibers that can improve the preformance flame retandant in furniture.

## Nanomaterial Applications using Nanoparticles

Applications being developed for nanoparticles include deliver chemotherapy drugs directly to cancer tumors, resetting the immune system to prevent autoimmune diseases, delivering drugs to damaged regions of arteries to fight cardiovascular disease, create photocatalysts that produce hydrogen from water, reduce the cost of producing fuel cells and solar cells, clean up oil spills, water pollution and air pollution.

## Nanomaterial Applications using Nanowires

Applications being developed for carbon nanotubes include using zinc oxide nanowires in a flexible solar cell, silver chloride nanowires to decompose organic molecules in polluted water, using nanowires made from iron and nickel to make dense computer memory - called "race track memory.

# Sources of Nanomaterials

Nanoparticles, one of the "building blocks" of nanotechnology are all around us right now, and have been all around us throughout human history. They were with us when human beings began making their first tools, and they are present in products we buy at the grocery store every day. They largely flew under the radar until electron microscopes become commonplace several decades ago, but now, the more we turn our microscopes on everyday objects, the more nanoparticles we seem to find.

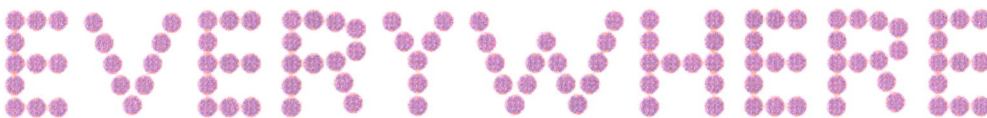

EVERYWHERE

Even the most seemingly mundane objects can give rise to nanoparticles; detecting them is simply a matter of being able to look closely enough to see them (no simple matter for such small materials). You could find nanoparticles in your jewelry box or the drawer with your family's fanciest silverware. I got to see this first hand while I was working in the Hutchison lab at the University of Oregon several years ago. Some of my colleagues were trying to understand why silver nanoparticles change size and shape so rapidly, even when they are just left in storage on the shelf. Because they saw such rapid changes in the size and shape of silver nanoparticles, they thought to look and see if large every day pieces of silver and copper (Sterling silver forks, earrings, and wires) might give off nanoparticles. To test this, they simply left the fork (or any of the other items) on an electron microscopy grid for several hours, then took the fork away, and had a peek at what it had left behind. Surprisingly, they found that the silver and copper items had left silver and copper nanoparticles behind all over the grid; a most elegant demonstration that human beings can come into contact with a variety of nanoparticles, even in our own homes. Forks and earrings are merely the tip of the iceberg, though. Wherever we go during our day-to-day routine we can encounter nanoparticles (both synthetic and natural).

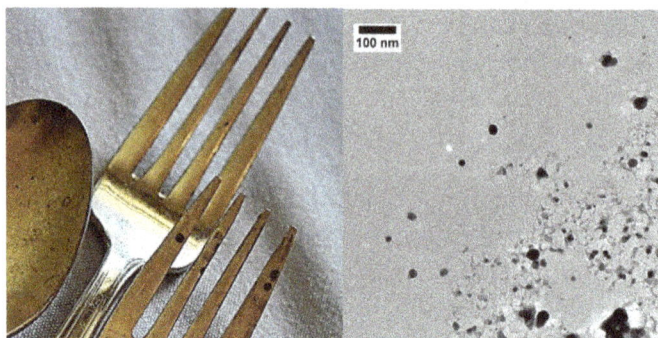

An experiment you could do at home (if you have your very own electron microscope).
Sterling silver forks release small quantities of silver nanoparticles into the surrounding environment.

It seems weird to say, but (as the fork study shows) nanoparticles are actually a fairly common type of material in many different environments, and they can pass by almost undetected unless you are looking for them. In fact, many kinds of physical and chemical processes (both human activities and natural processes) produce nanoparticles. Naturally occurring nanoparticles can be found in volcanic ash, ocean spray, fine sand and dust, and even biological matter (e.g. viruses). Synthetic nanoparticles are equally, if not more diverse than their naturally occurring counterparts.

Synthetic nanoparticles (sometimes called anthropogenic nanoparticles) fall into two general categories: "incidental" and "engineered" nanoparticles. Incidental nanoparticles are the byproducts of human activities, generally have poorly controlled sizes and shapes, and may be made of a hodgepodge of different elements. Many of the processes that generate incidental nanoparticles are common every day activities: running diesel engines, large-scale mining, and even starting a fire.

A few sources of incidental nanoparticles

Engineered nanoparticles on the other hand, have been specifically designed and deliberately synthesized by human beings. Not surprisingly, they have very precisely controlled sizes, shapes, and compositions. They may even contain "layers" with different chemical compositions (e.g. a core made out of gold, covered in a shell of silica, and coated with specifically chosen antibodies). Although engineered nanoparticles get more sophisticated with each passing year, simple engineered nanoparticles can be created by relatively simple chemical reactions that have been within the scope of chemists and alchemists for many centuries. This means that long before people could "see" a nanoparticle through an electron microscope, human beings were both deliberately and accidentally generating a wide variety of these materials.

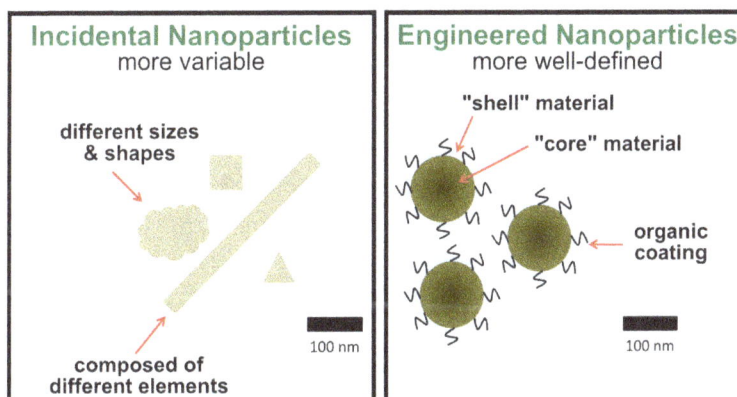

There are two types of synthetic nanoparticles: "incidental" (byproducts of other human activities) and engineered (made to order).

It's hard to say when human beings started making incidental nanoparticles, but probably as soon as people started taming fire—you can find small nano-scale particles of soot in the embers and smoke. Certainly, by the Bronze Age, incidental copper nanoparticles would have been prevalent in human civilizations. The earliest engineered nanoparticles are often attributed to the ancient Romans, Egyptians, and Chinese. Despite not having any idea about the scientific implications of what they were making, they were successfully able to prepare nanoparticle solutions of gold and other precious metals with reasonably precise control over particle size and composition. In fact, some of the methods used in these ancient cultures to make nanoparticles were very similar to the way many metal nanoparticles are still synthesized in laboratories today. These antiquarian gold nanoparticle solutions were vibrantly colored (generally reds and purples) and could be impregnated into glass to make stained glass and jewelry. In addition, many people ingested these colloidal gold and silver solutions as health tonics; specifically to treat high fevers and syphilis. Though it's unlikely that drinking metal nanoparticles has any truly positive aspects for your health, some people still swear by this folk remedy, with often unappetizing, though not necessarily lethal, results.

Roman colloidal gold. (L) The Lycurgus cup. Gold and silver nanoparticles in the glass make for some fantastic and very unique color effects. (R) A solution of gold nanoparticles in water

As human society has advanced technologically, it is likely that exposure to all kinds of synthetic nanoparticles has dramatically increased and will continue to increase. The advent of the industrial revolution provided an opportunity for human beings to be exposed to the byproducts of mining and fossil fuel burning on a scale never before seen. Indeed, the prevalence of black lung in miners for the past several hundred years can be linked to the respiration of ultrafine (aka nano) particles, and even today, people living near high-traffic areas/large-scale manufacturing operations often experience a much higher incidence of chronic respiratory problems that are beginning to be linked with nanoparticles found in diesel exhaust. Ever since researchers began to consider nanotechnology to be a unique research field, however, we have also seen a sharp rise in consumer products that deliberately include synthetic nanoparticles. These nano-enabled products range from cosmetics to sporting goods and the nanoparticless they contain serve a variety of functions: everything from acting as UV-blocking agents to making plastics that are extremely light, but stronger than steel. Some of these nanoparticle-enabled products actually made it onto the market before people recognized that their benefits were specifically related to the nanoparticles that they contained (e.g. silver impregnated wound dressings). The Woodrow Wilson Institute supports a database of ordinary consumer products that contain nanoparticles, and they now list more than 800 registered nanoparticle-enabled products available to consumers.

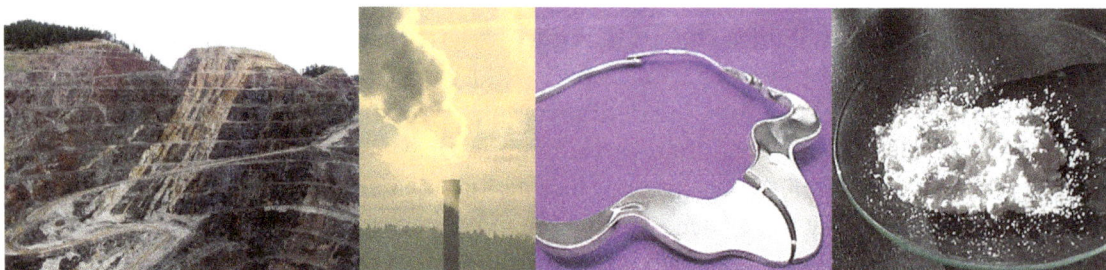

Where can we find nanoparticles in our everyday world? (L-R) Industrial processes like large-scale mining and fossil fuel burning can release nanoparticles, your best silverware and jewelry can give off small quantities of silver or copper nanoparticles, powdered titanium dioxide, like you might find in sunscreen.

Perhaps the easiest place to find nanoparticles in your very own home is in health care products and cosmetics. A number of companies have, for at least the past decade, sold sun block that contains combinations of zinc oxide (ZnO) and titanium dioxide ($TiO_2$) nanoparticles because these materials absorb UV-light very strongly, preventing your skin from feeling the unwanted effects of ultra-violet exposure. The smaller these particles are, the easier it is to keep them evenly mixed in the liquid portion of the sunblock. This means you are less likely to notice the chalky bits of semiconductor on your skin and the less likely you are to look like you are covered in chalk. Therefore, using $TiO_2$ and ZnO nanoparticles is an excellent way to ensure comfortable and uniform sun protection.

Clearly, it's not that hard to find nanoparticles anywhere we go in our everyday lives. We inhale them in tiny amounts as we go about our day-to-day activities, we buy products that use nanoparticles to protect our health, and even objects as mundane as a fork leave behind nanoparticles as they sit unused in drawers. Until recently, we haven't really noticed how often we are exposed to nanoparticles, because they can be extremely difficult to detect (unless you are specifically looking for them with the right equipment). One question this raises is: "Is this ever-increasing level of nanoparticle exposure good for us?" Unfortunately, we have no easy answer to that. As it happens, some of the nanoparticles that you are exposed to everyday that can be harmful to you (particularly the nano-sized particles that can be found in certain types of pollution) but others have already become our go-to weapons to prevent disease (whether that disease is skin cancer or bacterial infection).

## References

- What-is-nanoscience: tmi.utexas.edu, Retrieved 28 May 2018

- Whatisnano, nanoscience: tcd.ie, Retrieved 11 March 2018

- Nanomaterials, opinions-layman, scientific-committees: europa.eu, Retrieved 11 March 2018

- Nanomaterials: understandingnano.com, Retrieved 12 July 2018

- Nanoparticles-are-all-around-us: sustainable-nano.com, Retrieved 22 June 2018

# 2

# Nanoparticles

Nanoparticles are particles, which are between 1 and 100 nm in size that are surrounded by an interfacial layer. The characterization of the chemical and physical properties of nanoparticles is vital for nanotoxicology studies, exposure assessment and manufacturing process control. The topics elucidated in this chapter cover some of the important aspects of nanoparticles, such as the production and characterization of nanoparticles, magnetic and ferrite nanoparticles.

Nanoparticle is an ultrafine unit with dimensions measured in nanometers (nm; 1 nm = $10^{-9}$ meter). Nanoparticles exist in the natural world and are also created as a result of human activities. Because of their submicroscopic size, they have unique material characteristics, and manufactured nanoparticles may find practical applications in a variety of areas, including medicine, engineering, catalysis, and environmental remediation.

## Properties of Nanoparticles

In 2008 the International Organization for Standardization (ISO) defined a nanoparticle as a discrete nano-object where all three Cartesian dimensions are less than 100 nm. The ISO standard similarly defined two-dimensional nano-objects (i.e., nanodiscs and nanoplates) and one-dimensional nano-objects (i.e., nanofibres and nanotubes). But in 2011 the Commission of the European Union endorsed a more-technical but wider-ranging definition:

> "A natural, incidental or manufactured material containing particles, in an unbound state or as an aggregate or as an agglomerate and where, for 50% or more of the particles in the number size distribution, one or more external dimensions is in the size range 1 nm–100 nm."

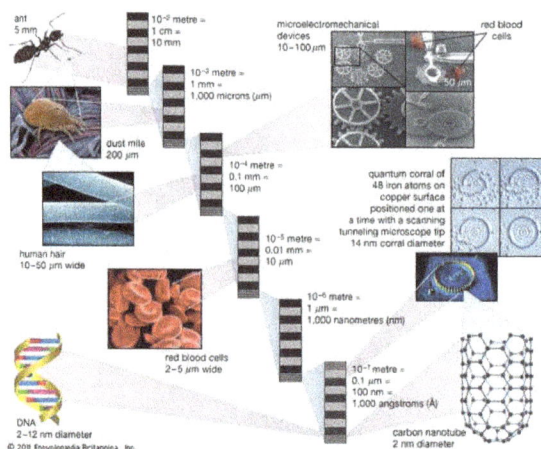

Examples from biological and mechanical realms illustrate various "orders of magnitude" (powers of 10), from $10^{-2}$ meter down to $10^{-7}$ meter.

Under that definition a nano-object needs only one of its characteristic dimensions to be in the range 1–100 nm to be classed as a nanoparticle, even if its other dimensions are outside that range. (The lower limit of 1 nm is used because atomic bond lengths are reached at 0.1 nm.)

That size range—from 1 to 100 nm—overlaps considerably with that previously assigned to the field of colloid science—from 1 to 1,000 nm—which is sometimes alternatively called the meso-scale. Thus, it is not uncommon to find literature that refers to nanoparticles and colloidal particles in equal terms. The difference is essentially semantic for particles below 100 nm in size.

There are three major physical properties of nanoparticles, and all are interrelated:

- They are highly mobile in the free state (e.g., in the absence of some other additional influence, a 10-nm-diameter nanosphere of silica has a sedimentation rate under gravity of 0.01 mm/day in water);

- They have enormous specific surface areas (e.g., a standard teaspoon, or about 6 ml, of 10-nm-diameter silica nanospheres has more surface area than a dozen doubles-sized tennis courts; 20 percent of all the atoms in each nanosphere will be located at the surface); and

- They may exhibit what are known as quantum effects. In addition, nanoparticles can be classified as hard (e.g., titania [titanium dioxide], silica [silica dioxide] particles, and fullerenes) or as soft (e.g., liposomes, vesicles, and nanodroplets). Thus, nanoparticles have a vast range of compositions, depending on the use or the product.

## Nanoparticle-based Technologies

In general, nanoparticle-based technologies center on opportunities for improving the efficiency, sustainability, and speed of already-existing processes. That is possible because, relative to the materials used traditionally for industrial processes (e.g., industrial catalysis), nanoparticle-based technologies use less material, a large proportion of which is already in a more "reactive" state. Other opportunities for nanoparticle-based technologies include the use of nanoscale zero-valent iron (NZVI) particles as a field-deployable means of remediating organochlorine compounds, such as polychlorinated biphenyls (PCBs), in the environment. NZVI particles are able to permeate into rock layers in the ground and thus can neutralize the reactivity of organochlorines in deep aquifers. Other applications of nanoparticles are those that stem from manipulating or arranging matter at the nanoscale to provide better coatings, composites, or additives and those that exploit the particles' quantum effects (e.g., quantum dots for imaging, nanowires for molecular electronics, and technologies for spintronics and molecular magnets).

Nanowires as seen by a field-emission microscope.

## Nanoparticle Applications in Materials

Many properties unique to nanoparticles are related specifically to the particles' size. It is therefore natural that efforts have been made to capture some of those properties by incorporating nanoparticles into composite materials. An example of how the unique properties of nanoparticles have been put to use in a nanocomposite material is the modern rubber tire, which typically is a composite of a rubber (an elastomer) and an inorganic filler (a reinforcing particle), such as carbon black or silica nanoparticles.

For most nanocomposite materials, the process of incorporating nanoparticles is not straightforward. Nanoparticles are notoriously prone to agglomeration, resulting in the formation of large clumps that are difficult to redisperse. In addition, nanoparticles do not always retain their unique size-related properties when they are incorporated into a composite material.

Despite the difficulties with manufacture, the use of nanomaterials grew markedly in the early 21st century, with especially rapid growth in the use of nanocomposites. Nanocomposites were employed in the development and design of new materials, serving, for example, as the building blocks for new dielectric (insulating) and magnetic materials. The following sections describe some of the many applications of nanoparticles and nanocomposites in materials.

### Polymers

Similar to the way in which carbon and silica nanoparticles have been used as fillers in rubber to improve the mechanical properties of tires, such particles and others, including nanoclays, have been incorporated into polymers to improve their strength and impact resistance. In the early 21st century, increasing use of non-petroleum-based polymers that were derived from natural sources drove the development of "all-natural" nanocomposite polymers. Such materials incorporate a biopolymer derived from an alginate (a carbohydrate found in the cell wall of brown algae), cellulose, or starch; the biopolymer is used in conjunction with a natural nanoclay or a filler derived from the shells of crustaceans. The materials are biodegradable and do not leave behind potentially harmful or non-natural residues.

### Food Packaging

Nanoparticles have been increasingly incorporated into food packaging to control the ambient atmosphere around food, keeping it fresh and safe from microbial contamination. Such composites use nanoflakes of clays and claylike particles, which slow down the ingress of moisture and reduce gas transport across the packaging film. It is also possible to incorporate nanoparticles with apparent antimicrobial effects (e.g., nanocopper or nanosilver) into such packaging. Nanoparticles that exhibit antimicrobial activity had also been incorporated into paints and coatings, making those products particularly useful for surfaces in hospitals and other medical facilities and in areas of food preparation.

### Flame Retardants

Nanoparticles were explored for their potential to replace additives based on flammable organic halogens and phosphorus in plastics and textiles. Studies had suggested that, in the event of a serious fire, products with nanoclays and hydroxide nanoparticles were associated with fewer emissions of harmful fumes than products containing certain other types of additives.

## Batteries and Super Capacitors

The ability to engineer nanocomposite materials to have very high internal surface areas for storing electrical charge in the form of small ions or electrons has made them especially valuable for use in batteries and super capacitors. Indeed, nanocomposite materials have been synthesized for various applications involving electrodes. Composite materials based on carbon nanotubes and layered-type materials, such as graphene, were also researched extensively, making their first appearances in commercial devices in the early 2000s.

## Nanoceramics

A long-term objective in materials science had been to transform ceramics that are brittle and prone to cracking into tougher, more resilient materials. By the early 21st century, researchers had achieved that goal by incorporating an effective blend of nanoparticles into ceramics materials. Other new ceramics materials that were under development included all-ceramic or polymer-ceramic blends, which combined the unique functional (e.g., electrical, magnetic, or mechanical) properties of a nanocomposite material with the properties of ceramics materials.

## Light Control

In the 1990s the development of blue light-emitting diodes (LEDs), which had the potential to produce white light at significantly reduced costs, inspired a revolution in lighting. Blue LEDs brought about a need for composite materials that could be used to coat the diodes to convert blue light into other wavelengths (such as red, yellow, or green) in order to achieve white light. One way of obtaining the desired light is by leveraging the size or quantum effect of small semiconducting particles. The application of such particles facilitated the development of nanocomposite polymers for greenhouse enclosures; the polymers optimize plant growth by effectively converting wavelengths of full-spectrum sunlight into the red and blue wavelengths used in photosynthesis. Light conversion in the above cases is achieved with submicron particles of inorganic phosphor materials incorporated into the polymer.

## Nanoparticle Applications in Medicine

The small size of nanoparticles is especially advantageous in medicine; nanoparticles can not only circulate widely throughout the body but also enter cells or be designed to bind to specific cells. Those properties have enabled new ways of enhancing images of organs as well as tumors and other diseased tissues in the body. They also have facilitated the development of new methods of delivering therapy, such as by providing local heating (hyperthermia), by blocking vasculature to diseased tissues and tumors, or by carrying payloads of drugs.

Magnetic nanoparticles have been used to replace radioactive technetium for tracking the spread of cancer along lymph nodes. The nanoparticles work by exploiting the change in contrast brought about by tiny particles of super paramagnetic iron oxide in magnetic resonance imaging (MRI). Such particles also can be used to kill tumors via hyperthermia, in which an alternating magnetic field causes them to heat and destroy tissue on a local scale.

Nanoparticles can be designed to enhance fluorescent imaging or to enhance images from positron emission tomography (PET) or ultrasound. Those methods typically require that the nanoparticle be able to recognize a particular cell or disease state. In theory, the same idea of targeting could be used in aiding the precise delivery of a drug to a given disease site. The drug could be carried via a nanocapsule or a liposome, or it could be carried in a porous nanosponge structure and then held by bonds at the targeted site, thereby allowing the slow release of drug. The development of nanoparticles to aid in the delivery of a drug to the brain via inhalation holds considerable promise for the treatment of neurological disorders such as Parkinson disease, Alzheimer disease, and multiple sclerosis.

Nanoparticles and nanofibres play an important part in the design and manufacture of novel scaffold structures for tissue and bone repair. The nanomaterials used in such scaffolds are biocompatible. For example, nanoparticles of calcium hydroxyapatite, a natural component of bone, used in combination with collagen or collagen substitutes could be used in future tissue-repair therapies.

Nanoparticles also have been used in the development of health-related products. For example, a sunscreen known as Optisol, invented at the University of Oxford in the 1990s, was designed with the objective of developing a safe sunscreen that was transparent in visible light but retained ultraviolet-blocking action on the skin. The ingredients traditionally used in sunscreens were based on large particles of either zinc oxide or titanium dioxide or contained an organic sunlight-absorbing compound. However, those materials were not satisfactory: zinc oxide and titanium dioxide are very potent photocatalysts, and in the presence of water and sunlight they generate free radicals, which have the potential to damage skin cells and DNA (deoxyribonucleic acid). Scientists proceeded to develop a nanoparticle form of titanium oxide that contained a small amount of manganese. Studies indicated that the nanoparticle-based sunscreen was safer than sunscreen products manufactured by using traditional materials. The improvement in safety was attributed to the introduction of manganese, which changed the semiconducting properties of the compound from n-type to p-type, thus shifting its Fermi level, or oxidation-reduction properties, and making the generation of free radicals less likely.

Treatments and diagnostic approaches based on the use of nanoparticles are expected to have important benefits for medicine in the future, but the use of nanoparticles also presents significant challenges, particularly regarding impacts on human health. For example, little is known about the fate of nanoparticles that are introduced into the body or whether they have undesirable effects on the body. Extensive clinical trials are needed in order to fully address concerns about the safety and effectiveness of nanoparticles used in medicine. There also are manufacturing problems to be overcome, such as the ability to produce nanoparticles under sterile conditions, which is required for medical applications.

## Manufacture of Nanoparticles

Nanoparticles are made by one of three routes: by comminution (the pulverization of materials), such as through industrial milling or natural weathering; by pyrolysis (incineration); or by sol-gel synthesis (the generation of inorganic materials from a colloidal suspension). Comminution is known as a top-down approach, whereas the sol-gel process is a bottom-up approach. Examples of those three processes (comminution, pyrolysis, and sol-gel synthesis) include the production of titania nanoparticles for sunscreens from the minerals anatase and rutile, the production of fullerenes or fumed silica, and the production of synthetic (or Stöber) silica, of other "engineered" oxide nanoparticles, and of quantum dots. For the generation of small nanoparticles, comminution is a very inefficient process.

## Detection, Characterization and Isolation

The detection and characterization of nanoparticles present scientists with particular challenges. Being of a size that is at least four to seven times smaller than the wavelength of light means that individual nanoparticles cannot be detected by the human eye, and they are observable under optical microscopes only in liquid samples under certain conditions. Thus, in general, specialized techniques are required to see them, and none of those approaches is currently field-deployable.

Techniques to detect and characterize nanoparticles fall into two categories: direct, or "real space," and indirect, or "reciprocal space." Direct techniques include transmission electron microscopy (TEM), scanning electron microscopy (SEM), and atomic force microscopy (AFM). Those techniques can image nanoparticles, directly measure sizes, and infer shape information, but they are limited to studying only a few particles at a time. There are also significant issues surrounding sample preparation for electron microscopy. In general, however, those techniques can be quite effective for obtaining basic information about a nanoparticle.

Indirect techniques use X-rays or neutron beams and obtain their information by mathematically analyzing the radiation scattered or diffracted by the nanoparticles. The techniques of greatest relevance to nanoscience are small-angle X-ray scattering (SAXS) and small-angle neutron scattering (SANS), along with their surface-specific analogues GISAXS and GISANS, where GI is "grazing incidence," and X-ray or neutron reflectometry (XR/NR). The advantage of those techniques is that they are able to simultaneously sample and average very large numbers of nanoparticles and often do not require any particular sample preparation. Indirect techniques have many applications. For example, in studies of nanoparticles in raw sewage, scientists used SANS measurements, in which neutrons readily penetrated the turbid sewage and scattered strongly from the nanoparticles, to follow the aggregation behavior of the particles over time.

The isolation of nanoparticles from colloidal and larger matter involves specialized techniques, such as ultra-centrifugation and field-flow fractionation. Such laboratory-based techniques are normally coupled to standard spectroscopic instrumentation to enable particular types of chemical characterization.

## Nanoparticles in the Environment

Nanoparticles occur naturally in the environment in large volumes. For example, the sea emits an aerosol of salt that ends up floating around in the atmosphere in a range of sizes, from a few nanometers upward, and smoke from volcanoes and fires contains a huge variety of nanoparticles, many of which could be classified as dangerous to human health. Dust from deserts, fields, and so on also has a range of sizes and types of particles, and even trees emit nanoparticles of hydrocarbon compounds such as terpenes (which produce the familiar blue haze seen in forests, from which the Great Smoky Mountains in the United States get their name).

Human-made (anthropogenic) nanoparticles are emitted by large industrial processes, and in modern life it is particles from power stations and from jet aircraft and other vehicles (namely, those powered by internal-combustion engines; car tires are also a factor) that constitute the major fraction of nanoparticle emissions. Types of nanoparticles that are emitted include partially burned hydrocarbons (in soot), ceria (cerium oxide; from vehicle exhaust catalysts), metallic dust (from brake linings), calcium carbonate (in engine lubricating oils), and silica (from car tires). Other sources of nanoparticles to the environment include the semiconductor industry, domestic and industrial wastewater discharges, the health care industry, and the photographic industry. However, all those emission levels are still considered to be lower than the levels of nanoparticles produced through natural processes. Indeed, recent human-made particles contribute only a small amount to air and water pollution.

Understanding the relationship between nanoparticles and the environment forms an important area of research. There are several mechanisms by which nanoparticles are believed to affect the environment negatively. Two scenarios that are under investigation are the possibilities (1) that the mobility and sorptive capacity of nanoparticles (natural or human-made) make them potent vectors (carriers) in the transport of chemical pollutants (e.g., phosphorus from sewage and agriculture), particularly in rivers and lakes, and (2) that some nanoparticles are able to reduce the functioning of (and may even disrupt or kill) naturally occurring microbial communities, as well as microbial communities that are employed in industrial processes (e.g., those that are used in sanitation processes, including sewage treatment).

Nanoparticles also can have beneficial impacts on the environment and appear to contribute to natural processes. Thus, in addition to the potential use of nanoparticles to remove chemical contaminants from the environment, scientists are investigating how nanoparticles interact with all life-forms—from fungi to microbes, algae, plants, and higher-order animals. That type of study is essential not only to improving scientists' knowledge of nanoparticles but also to gaining a more complete understanding of life on Earth, since the soil is naturally full of nanoparticles, in a richly diverse environment.

## Health Effects of Nanoparticles

Humans have evolved to cope with most naturally occurring nanoparticles. However, some nanoparticles, generated as a result of certain human activities such as tobacco smoking and fires, account for many premature deaths as a result of lung damage. For example, fires from the types of cooking stoves used in developing countries are known to emit fine particles and lead to early mortality, especially among women who routinely work near the stoves.

Laboratory and clinical investigations of the effects of nanoparticles on health have been somewhat controversial and remain largely inconclusive. Most studies in animals have involved nanoparticle inhalation, and the dosages have been very large. The results of those studies have indicated that large quantities of nanoparticles can cause cellular damage in the lungs, with lung cells absorbing the particles and becoming damaged or undergoing genetic mutation. However, the health effects of typical exposure levels—those that are encountered by most persons during daily activities—remain unknown. Nonetheless, there is a general awareness of the problems that might occur upon excess exposure to nanoparticles, and, thus, most manufacturers of such particles take serious precautions to avoid exposure of their workers. Efforts have been made to educate the public in the use of nanoparticle-containing products. The existence of pressure groups has also helped to ensure nanoparticle safety compliance among manufacturers. However, nanoparticles offer tremendous potential for new or improved forms of health care treatment. That has spawned a new field of science called nanomedicine.

# Characterization of Nanoparticles

Nanoparticles are entities some billionths of a meter in size. The formal definition of a nanoparticle is a "nano-object with all three external dimensions in the nanoscale", although in practice the term is often used to refer to particles larger than 100 nm. The reason for this is that the behavior of nanoparticles and the applicability of measurement techniques vary with size and environment, to the extent that 500 nm particles can either be considered very large or very small depending on the frame of reference.

Nanoparticles can be both natural and man-made entities, and are widely found in the environment as well as the laboratory. Their origins and properties are highly varied, making their study a rich branch of analytical science.

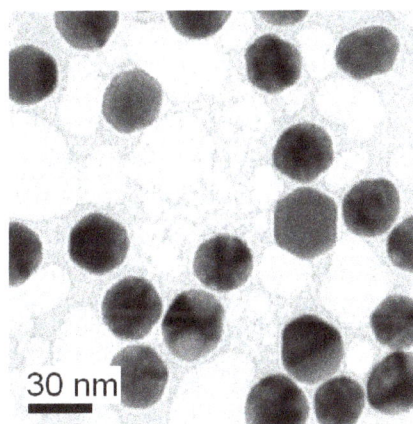

Figure: Transmission electron micrograph of a collection of gold nanoparticles.

The properties of nanoparticles often bridge the microscopic and macroscopic regimes, meaning that conventional theories do not necessarily allow us to predict their behavior. It is this uncertainty that lies at the heart of concerns surrounding the health and environmental impact of nanoparticles, but also of the excitement around opportunities for their application in new areas of science and technology. Therefore, it is important to have robust analytical approaches for characterizing nanoparticles, to maximize our benefit from them whilst mitigating their impact.

## Material Properties

Chemical properties of interest for those studying nanoparticles include total chemical composition, mixing state (internal/external), surface composition, electrochemistry and oxidation state. Physical properties of interest include number and mass concentration, size, surface area, total mass, morphology and optical properties. Because of their very high surface area to mass ratio, and high surface curvature, nanoparticles may be particularly chemically active.

Table: Comparison of the properties of nominally spherical nanoparticles and their likely penetration into the human respiratory system. Particle size has a dramatic effect on the physical properties of a collection of particles. Mass-based measurements are heavily weighted towards the largest particles, whereas smaller particles have a much larger surface area per unit mass.

| Particle Diameter (nm) | Relative Mass per Particle | Relative Surface Area per Unit Mass | Respiratory Penetration |
|---|---|---|---|
| 10 000 | 8 000 000 | 1 | Nasal Passage |
| 2 500 | 125 000 | 16 | Trachea, Bronchi |
| 1000 | 8 000 | 100 | Alveoli |
| 50 | 1 | 40 000 | Bloodstream? |

Nanoparticle characterization methods are required to cover a range of requirements, from long term environmental monitoring campaigns over entire continents where only basic properties are measured, to one-off laboratory measurements on a specially prepared samples where a full chemical and physical analysis is performed.

## Characterization Techniques

Characterization techniques can be subdivided by both general measure and the phase in which the nanoparticles reside. Measurements of each type present their own difficulties and often have subtly different interpretations. Moreover, comparison of results between phases is very difficult, and matrix effects can be significant due to the high surface area to mass ratio of nanoparticles. The techniques presented below give a general overview of common measurements made on nanoparticles for a range of applications.

## Solid Phase

Nanoparticles in the solid phase exist either as a powder or encapsulated in a solid medium. The former can take several forms including loose powders and wet or dry 'powder cakes' for convenience of handling. As such, any analysis must take into account how the particles will eventually be used because this will affect their final agglomeration state and other properties.

- Size: There are many methods of measuring particle size (or, more correctly, size distribution) but electron microscopy is widely used, in conjunction with other measurements. Laser diffraction is a common technique for measuring bulk samples under ambient conditions and powder X-ray diffraction may also be used by examining peak broadening.

- Surface Area: The most common technique is the nitrogen adsorption technique based on the BET isotherm, and is routinely carried out in many laboratories.

- Pour Density: This requires the weighing of a known volume of freshly poured powder.

- Composition: Suitable surface techniques include X-ray photoelectron spectroscopy (and other X-ray spectroscopy methods) and secondary ion mass spectrometry. Bulk techniques generally use digestion followed by conventional wet chemical analyses such as mass spectrometry, atomic emission spectroscopy and ion chromatography.

- Crystallography: Powder X-ray or neutron diffraction may be readily used to determine the crystal structure of simple lattice structures. It can also be applied to crystalline organic solids, but the data analysis is much more challenging.

- 'Dustiness': This is the propensity of a powder to become aerosolized by mechanical agitation. The dustiness of a particular sample is very dependent on its moisture content and static electrical properties.

- Morphology: Particle shape and aspect ratio are most readily determined by image analysis of electron micrographs.

## Liquid Phase

- Nanoparticles suspended in the liquid phase are often mixed with surfactants to moderate their agglomeration state. Furthermore, a range of chemical or even biological species may be present and affect the results obtained, especially in samples taken from the environment.

- Therefore, care must be taken to prepare the sample to prevent unwanted matrix effects or changes to the sample.

- Size: Dynamic light scattering (photon correlation spectroscopy) and centrifugation are commonly used techniques. Other methods include image-tracking instruments.

- Surface area: Simple titrations may be used to estimate surface area but are very labor intensive. NMR experiments may also be used and dedicated instrumentation has recently been developed.

- Zeta potential: This gives information relating to the stability of dispersions. It cannot be measured directly but is often measured indirectly using an electrophoretic method.

- Composition: Chemical digestion of the particles allows a range of mass spectroscopy and chromatography methods to be used. Interference from the matrix needs to be carefully managed.

- Morphology: It is difficult to measure morphology of particles moving freely in a fluid. Deposition onto a surface for electron microscopy is most commonly used.

## Gas Phase (Aerosols)

Nanoparticles in the gas phase may be monitored using a range of commercially available and relatively low cost equipment. Generally, these instruments are quite robust and can be used for prolonged periods with little attention. They are also generally resistant to matrix effects.

However, relative humidity and volatile organic species can sometimes affect measurements. Unlike the condensed phases, aerosols cannot be stored for later analysis and so reproducible sampling is very important. A range of inlets has been designed for different applications to reduce inconsistencies in this respect.

- 'Concentration': Usually expressed as a number concentration or mass concentration. The latter will strongly weight the distribution curve in favor of larger particles - a single 10 μm diameter particle weighs the same as 1 million 100 nm particles.

- Size: There are many methods for measuring particle size but comparability between them is a problem. They include optical and aerodynamic methods, but these give no information about variations in morphology.

- Surface area: There are few routine surface area techniques available. The most common involves charging the aerosol using a corona discharge and measuring the charge concentration. Such methods are usually calibrated against size distribution measurements.

- Charge: Collisions of air ions with particles result in a steady state charge distribution. This distribution is Boltzmann-like, with small particles being much less likely to carry any charge than larger particles. This distribution is usually measured in the laboratory and subsequently utilized by other measurement techniques.

- Morphology: There are few instruments available for measuring morphology. The most common technique is to capture particles either electrostatically or by filtration for subsequent imaging using electron microscopy.

- Composition: Measuring aerosol composition is very challenging owing to the small amounts of matter present. Most speciation methods require the particles to be collected and then subjected to spectrometric or wet chemical techniques.

## Methods used for Characterization

## Nanomaterial Characterization by Microscopy

Optical microscopes are generally used for observing micron level materials with reasonable resolution. Further magnification cannot be achieved through optical microscopes due to aberrations and limit in wavelength of light. Hence, the imaging techniques such as scanning electron microscopy (SEM), transmission electron microscopy (TEM/HRTEM), scanning tunneling microscopy (STM), atomic force microscopy (AFM), etc. have been developed to observe the submicron size materials. Though the principles of all the techniques are different but one common thing is that they produce a highly magnified image of the surface or the bulk of the sample. Nanomaterials can only be observed through these imaging techniques as human eye as well as optical microscope cannot be used to see dimensions at nano-level.

## Scanning Electron Microscopy (SEM)

The scanning electron microscope is an electron microscope that images the sample surface by scanning it with a high energy beam of electrons. Conventional light microscopes use a series of glass lenses to bend light waves and create a magnified image while the scanning electron microscope creates the magnified images by using electrons instead of light waves.

## Basic Principle

When the beam of electrons strikes the surface of the specimen and interacts with the atoms of the sample, signals in the form of secondary electrons, back scattered electrons and characteristic X-rays are generated that contain information about the sample's surface topography, composition, etc.

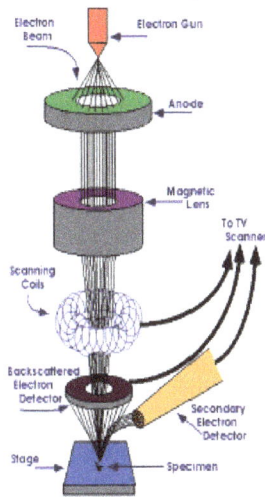

Schematic diagram of SEM1

The SEM can produce very high-resolution images of a sample surface, revealing details about 1-5 nm in size in its primary detection mode i.e. secondary electron imaging. Characteristic X-rays are the second most common imaging mode for an SEM. These characteristic X-rays are used to identify the elemental composition of the sample by a technique known as energy dispersive X- ray (EDX). Back-scattered electrons (BSE) that come from the sample may also be used to form an image. BSE images are often used in analytical SEM along with the spectra made from the characteristic X-rays as clues to the elemental composition of the sample.

In a typical SEM, the beam passes through pairs of scanning coils or pairs of deflector plates in the electron column to the final lens, which deflect the beam horizontally and vertically so that it scans in a raster fashion over a rectangular area of the sample surface. Electronic devices are used to detect and amplify the signals and display them as an image on a cathode ray tube in which the raster scanning is synchronized with that of the microscope. The image displayed is therefore a distribution map of the intensity of the signal being emitted from the scanned area of the specimen.

SEM requires that the specimens should be conductive for the electron beam to scan the surface and that the electrons have a path to ground for conventional imaging. Non-conductive solid specimens are generally coated with a layer of conductive material by low vacuum sputter coating or high vacuum evaporation. This is done to prevent the accumulation of static electric charge on the specimen during electron irradiation. Non-conducting specimens may also be imaged uncoated using specialized SEM instrumentation such as the "Environmental SEM" or in field emission gun SEM operated at low voltage, high vacuum or at low vacuum, high voltage.

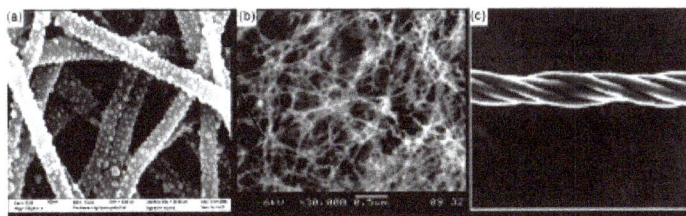

Figure: Electrospun nylon 6 nanofibres with surface bound silver nanoparticles,
(b) peptide nanofibre scaffold for tissue engineering, and (c) SEM image of plied CNT yarn

## Energy Dispersive X-ray Analysis (EDX)

Energy dispersive X-ray analysis is a technique to analyze near surface elements and estimate their proportion at different position, thus giving an overall mapping of the sample.

## Basic Principle

This technique is used in conjunction with SEM. An electron beam strikes the surface of a conducting sample. The energy of the beam is typically in the range 10-20keV. This causes X-rays to be emitted from the material. The energy of the X-rays emitted depends on the material under examination. The X- rays are generated in a region about 2 microns in depth, and thus EDX is not truly a surface science technique. By moving the electron beam across the material an image of each element in the sample can be obtained. Due to the low X-ray intensity, images usually take a number of hours to acquire.

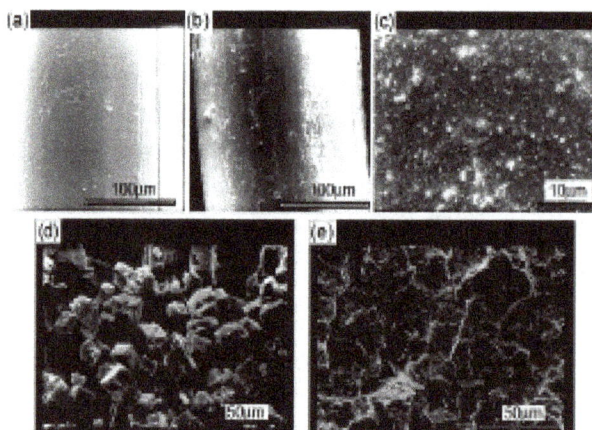

Figure: SEM surface images of (a) PP, (b & c) PP/clay nanocomposite filament, (d) POSS nanofillers, and (e) cross sectional view of HDPE/POSS nanocomposite fibre

## Transmission Electron Microscopy (TEM)

Transmission electron microscopy is a microscopy technique whereby a beam of electrons is transmitted through an ultra-thin specimen and interacts as passes through the sample. An image is formed from the electrons transmitted through the specimen, magnified and focused by an objective lens and appears on an imaging screen.

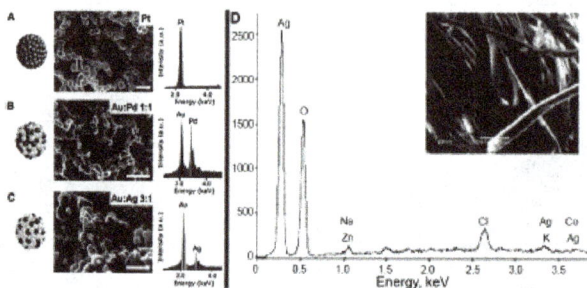

Figure: SEM images and EDX spectra of nanoporous materials made of (a) pure platinum, (b) 1:1 gold-palladium, (c) 3:1 gold- silver and (d) cotton cloth with silver nanoparticles

## Basic Principle

The contrast in a TEM image is not like the contrast in a light microscope image. In TEM, the crystalline sample interacts with the electron beam mostly by diffraction rather than by absorption. The intensity of the diffraction depends on the orientation of the planes of atoms in a crystal relative to the electron beam; at certain angles the electron beam is diffracted strongly from the axis of the incoming beam, while at other angles the beam is largely transmitted. Modern TEMs are equipped with specimen holders that allow to tilt the specimen to a range of angles in order to obtain specific diffraction conditions. Therefore, a high contrast image can be formed by blocking electrons deflected away from the optical axis of the microscope by placing the aperture to allow only unscattered electrons through. This produces a variation in the electron intensity that reveals information on the crystal structure. This technique, particularly sensitive to extended crystal lattice defects, is known as 'bright field' or 'light field'. It is also possible to produce an image from electrons deflected by a particular crystal plane which is known as a dark field image.

The specimens must be prepared as a thin foil so that the electron beam can penetrate. Materials that have dimensions small enough to be electron transparent, such as powders or nanotubes, can be quickly produced by the deposition of a dilute sample containing the specimen onto support grids. As polymeric nanocomposites or the textile samples are not as hard as metals, they are cut into thin films (< 100 nm) using ultra-microtome with diamond knife at cryogenic condition (in liquid nitrogen).

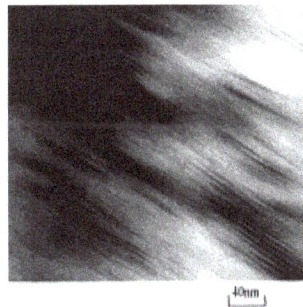

Figure: TEM images of PP/MMT nanocomposites

## High Resolution Transmission Electron Microscopy (HRTEM)

High resolution transmission electron microscopy is an imaging mode of the transmission electron microscope that allows the imaging of the crystallographic structure of a sample at an atomic scale.

## Basic Principle

As opposed to conventional microscopy, HRTEM does not use absorption by the sample for image formation, but the contrast arises from the interference in the image plane of the electron wave with itself. Each imaging electron interacts independently with the sample. As a result of the interaction with the sample, the electron wave passes through the imaging system of the microscope where it undergoes further phase change and interferes as the image wave in the imaging plane. It is important to realize that the recorded image is not a direct representation of the samples crystallographic structure.

## Atomic Force Microscope (AFM)

The atomic force microscope is ideal for quantitatively measuring the nanometer scale surface roughness and for visualizing the surface nano-texture on many types of material surfaces including polymer nanocomposites and nanofinished or nanocoated textiles. Advantages of the AFM for such applications are derived from the fact that the AFM is non- destructive technique and it has a very high three dimensional spatial resolution.

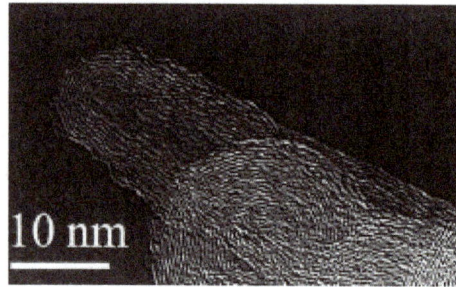

## Basic Principle

The basic principle and applications of atomic force microscopy have been the subject of a number of excellent reviews. In atomic force microscopy a probe consisting of a sharp tip (nominal tip radius is in the order of 10 nm) located near the end of a cantilever beam is raster scanned across the surface of a specimen using piezoelectric scanners. Changes in the tip specimen interaction are often monitored using an optical lever detection system, in which a laser is reflected off of the cantilever and onto a position- sensitive photodiode. During scanning, a particular operating parameter is maintained at a constant level, and images are generated through a feedback loop between the optical detection system and the piezoelectric scanners. There are three scan modes for AFM, namely contact mode, non-contact mode and tapping mode.

In contact mode, the tip scans the specimen in close contact with the surface of the material. The repulsive force on the tip is set by pushing the cantilever against the specimen's surface with a piezoelectric positioning element.

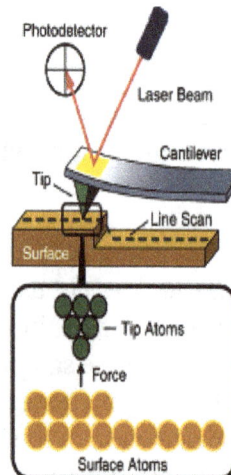

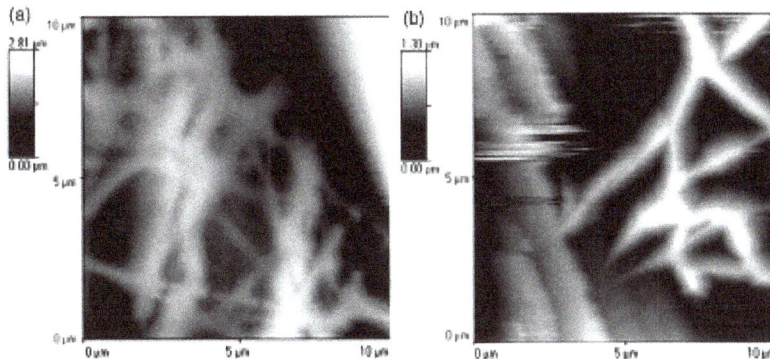

The deflection of the cantilever is measured and the AFM images are created. In non- contact mode, the scanning tip hovers about 50–150 Å above the specimen's surface. The attractive forces acting between the tip and the specimen are measured, and topographic images are constructed by scanning the tip above the surface. Tapping mode imaging is implemented in ambient air by oscillating the cantilever assembly at its resonant frequency (often hundreds of kilohertz) using a piezoelectric crystal. The piezo motion causes the cantilever to oscillate when the tip is not in contact with the surface of a material. The oscillating tip is then moved towards the surface until it begins to tap the surface. During scanning, the vertically oscillating tip alternately contacts the surface and lifts off, generally at a frequency of 50,000–500,000 cycles/s. As the oscillating cantilever begins to intermittently contact the surface, the cantilever oscillation is reduced due to energy loss caused by the tip contacting the surface. The reduction in oscillation amplitude is used to measure the surface characteristics.

Figure: 3D views of non-contact mode AFM images of PET textile surface (Scan area 1 μm × 1 μm)
(a) untreated surface, (b) 60s plasma treated surface, and (c) 120s plasma treated surface

## Scanning Tunneling Microscopy (STM)

Scanning tunneling microscopy is an instrument for producing surface images with atomic-scale lateral resolution, in which a fine probe tip is scanned over the surface of a conducting specimen, with the help of a piezoelectric crystal at a distance of 0.5–1 nm, and the resulting tunneling current or the position of the tip required to maintain a constant tunneling current is monitored.

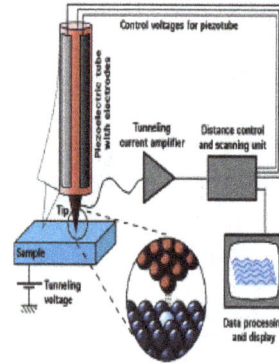

Figure: Schematic view of an STM

## Basic Principle

The principle of STM is based on the concept of quantum tunneling. When a conducting tip is brought very near to a metallic or semi-conducting surface, a bias between the two can allow electrons to tunnel through the vacuum between them. For low voltages, this tunneling current is a function of the local density of states at the Fermi level of the sample. Variations in current as the probe passes over the surface are translated into an image. For STM, good resolution is considered to be 0.1 nm lateral resolution and 0.01 nm depth resolution. They normally generate images by holding the current between the tip of the electrode and the specimen at some constant value by using a piezoelectric crystal to adjust the distance between the tip and the specimen surface, while the tip is piezoelectrically scanned in a raster pattern over the region of specimen surface being imaged by holding the force, rather than the electric current, between tip and specimen at a set-point value. Atomic force microscopes similarly allow the exploration of nonconducting specimens. In either case, when the height of the tip is plotted as a function of its lateral position over the specimen, an image that looks very much like the surface topography results. The STM can be used not only in ultra-high vacuum but also in air and various other liquid or gas, at ambient and wide range of temperatures. STM can be a challenging technique, as it requires extremely clean surfaces and sharp tips.

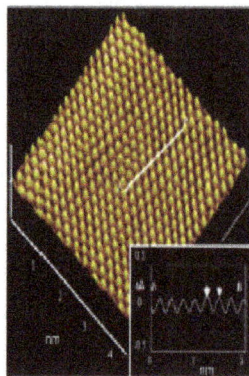

Figure: Highly oriented pyrolytic graphite sheet under STM

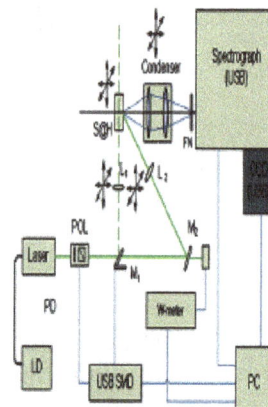

Figure: Schematic diagram of Raman spectrometer

# Nanomaterials Characterization by Spectroscopy

## Raman Spectroscopy

Raman spectroscopy is a spectroscopic technique used in condensed matter physics and chemistry to study vibrational, rotational, and other low-frequency modes in a system. It relies on inelastic scattering, or Raman scattering of monochromatic laser light. The laser light interacts with phonons or other excitations in the system, resulting in the energy of the laser photons being shifted up or down. The shift in energy gives information about the phonon modes in the system.

## Basic Principle

The Raman effect occurs when light impinges upon a molecule, interacts with the electron cloud of the bonds of that molecule and incident photon excites one of the electrons into a virtual state. For the spontaneous Raman effect, the molecule will be excited from the ground state to a virtual energy state, and relax into a vibrational excited state, which generates stokes Raman scattering. If the molecule was already in an elevated vibrational energy state, the Raman scattering is then called anti-stokes Raman scattering. A molecular polarizability change or amount of deformation of the electron cloud, with respect to the vibrational coordinate is required for the molecule to exhibit the Raman effect. The amount of the polarizability change will determine the Raman scattering intensity, whereas the Raman shift is equal to the vibrational level that is involved.

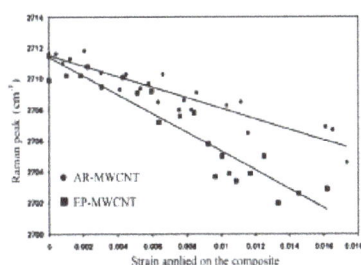

Figure: Shift in the Raman peak as a function of applied strain

## Ultraviolet-visible (UV-VIS) Spectroscopy

Ultraviolet spectrophotometers consist of a light source, reference and sample beams, a monochromator and a detector. The ultraviolet spectrum for a compound is obtained by exposing a sample of the compound to ultraviolet light from a light source, such as a Xenon lamp.

## Basic Principle

The reference beam in the spectrophotometer travels from the light source to the detector without interacting with the sample. The sample beam interacts with the sample exposing it to ultraviolet light of continuously changing wavelength. When the emitted wavelength corresponds to the energy level which promotes an electron to a higher molecular orbital, energy is absorbed. The detector records the ratio between reference and sample beam intensities ($I_o/I$). The computer determines at what wavelength the sample absorbed a large amount of ultraviolet light by scanning for the largest gap between the two beams. When a large gap between intensities is found, where the sample beam intensity is significantly weaker than the reference beam, the computer plots this wavelength as having the highest ultraviolet light absorbance when it prepares the ultraviolet absorbance spectrum.

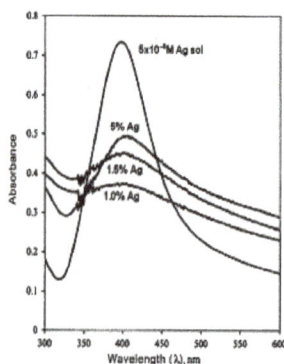

UV- Vis absorption spectrum of $5 \times 10^{-4}$ M Ag sol and spectra of suspensions of Ag/kaolinite samples at different silver contents (1, 1.5 and 5% Ag)

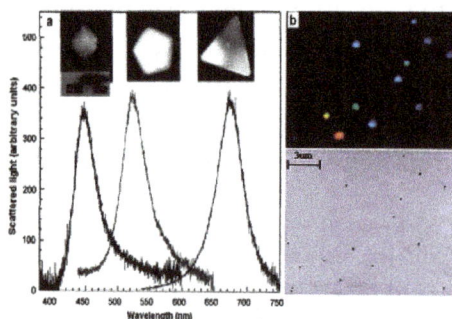

(a) Optical spectroscopy measurements of individual silver nanoparticles of different shapes and (b) colour image of a typical sample of silver nanoparticles as viewed under the dark field microscope (top picture), and a bright field TEM image of the same collection of silver nanoparticles (bottom picture)27

## Characterization of Nanomaterials by X-ray

### Wide Angle X-ray Diffraction

X-rays are electromagnetic radiation similar to light, but with a much shorter wavelength (few Angstrom). They are produced when electrically charged particles of sufficient energy are decelerated. In an X-ray tube, the high voltage maintained across the electrodes draws electrons toward a metal target (the anode). X-rays are produced at the point of impact, and radiate in all directions.

### Basic Principle

If an incident X-ray beam encounters a crystal lattice, general scattering occurs. Although most scattering interferes with itself and is eliminated (destructive interference), diffraction occurs when scattering in a certain direction is in phase with scattered rays from other atomic planes. Under this condition the reflections combine to form new enhanced wave fronts that mutually reinforce each other (constructive interference). The relation by which diffraction occurs is known as the Bragg's law or equation. As each crystalline material including the semi crystalline polymers as well as metal and metal oxide nanoparticles and layered silicate nanoclays have a characteristic atomic structure, it will diffract X-rays in a unique characteristic diffraction order or pattern.

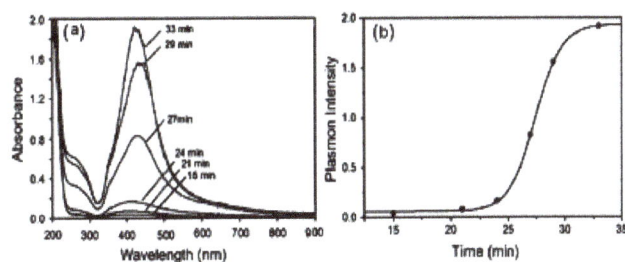

Figure: (a) UV–Vis absorption spectra taken from a reaction solution after the reactants had been mixed and sonicated in air at 27 °C for different periods, and (b) a plot of the intensity of the plasmon peak vs reaction time

## Ray Photoelectron Spectroscopy (XPS)

X-ray photoelectron spectroscopy is a quantitative spectroscopic surface chemical analysis technique used to estimate the empirical formula or elemental composition, chemical state and electronic state of the elements on the surface (up to 10 nm) of a material. XPS is also known as ESCA, an abbreviation of electron spectroscopy of chemical analysis.

## Basic Principle

X-ray irradiation of a material under ultra-high vacuum (UHV) leads to the emission of electrons from the core orbitals of the top 10 nm of the surface elements of the material being analyzed. Measurement of the kinetic energy (KE) and the number of electrons escaping from the surface of the material gives the XPS spectra. From the kinetic energy, the binding energy of the electrons to the surface atoms can be calculated. The binding energy of the electrons reflects the oxidation state of the specific surface elements. The number of electrons reflects the proportion of the specific elements on the surface.

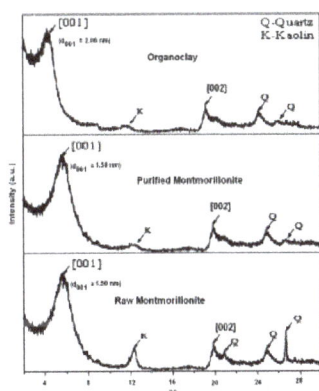

Figure: WAXD patterns for raw montmorillonite (MMT), purified montmorillonite and organoclay

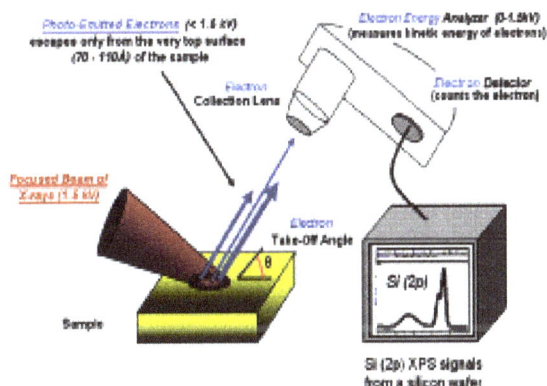

Figure: Schematic view of XPS

As the energy of a particular X-ray wavelength used to excite the electron from a core orbital is a known quantity, we can determine the electron binding energy (BE) of each of the emitted electrons by using the following equation that is based on the work of Ernest Rutherford:

$$E_{binding} = E_{photon}.E_{Kinetic} - \phi$$

where $E_{binding}$ is the energy of the electron emitted from one electron configuration within the atom; $E_{photon}$, the energy of the X-ray photons being used; $E_{kinetic}$, the kinetic energy of the emitted electron as measured by the instrument; and $\phi$ the work function of the spectrometer (not the material).

## Particle Size Analyzer

There are different techniques for the measurement of particle size and its distribution (PSD) such as sieve analysis, optical counting methods, electro resistance counting methods, sedimentation techniques, laser diffraction methods, dynamic light scattering method, acoustic spectroscopy, etc. Among them dynamic light scattering is mostly used for obtaining size distribution of nanoparticles.

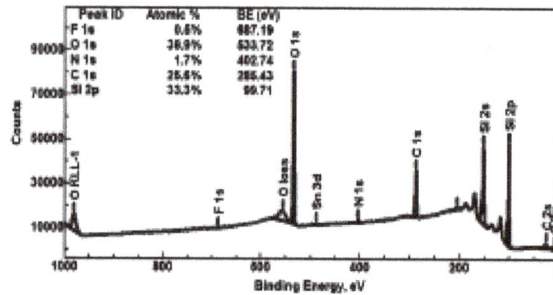

Figure: Wide XPS scan survey spectrum for all elements

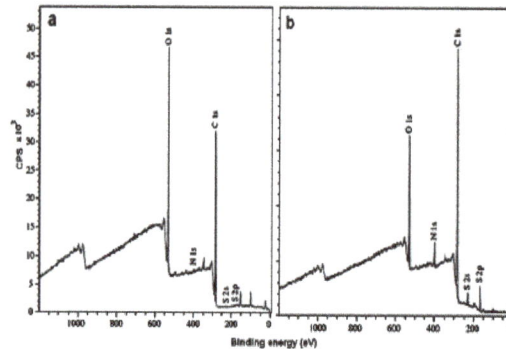

Figure: XPS spectra for (a) cationically charged woven cotton fabric and (b) cationically charged woven cotton fabric supporting 20 self-assembled layers of PSS/PAH

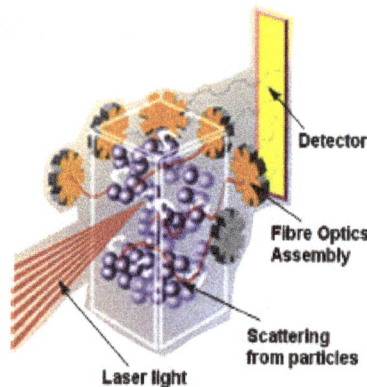

Figure: Schematic diagram of particle size analyzer

## Basic Principle of Dynamic Light Scattering (DLS)

Dynamic light scattering, sometimes referred to as photon correlation spectroscopy (PCS) or quasi- elastic light scattering (QELS) is a non- invasive, well-established technique for measuring the size of molecules and particles typically in the submicron region, and with the latest technology lower than 1 nanometer.

Particles, emulsions and molecules in suspension undergo Brownian motion. This is the motion induced by the bombardment of solvent molecules that themselves are moving due to their thermal energy. If the particles or molecules are illuminated with a laser, the intensity of the scattered light fluctuates at a rate that is dependent upon the size of the particles as smaller particles are "kicked" further by the solvent molecules and move more rapidly. Analysis of these intensity fluctuations yields the velocity of the Brownian motion and hence the particle size (radius $r_K$ ) using the Stokes-Einstein relationship, as shown below:

$$r_K = \frac{kT}{\pi \eta D}$$

Where, k is the Boltzmann's constant; $T$, the temperature in K; $\eta$ , the solvent viscosity; and D, the diffusion coefficient.

# Nanoparticle Production

Nanomaterials and/or nanoparticles are used in a broad spectrum of applications. Today they are contained in many products and used in various technologies. Most nanoproducts produced on an industrial scale are nanoparticles, although they also arise as byproducts in the manufacture of other materials. Most applications require a precisely defined, narrow range of particle sizes (monodispersity).

Specific synthesis processes are employed to produce the various nanoparticles, coatings, dispersions or composites.

Defined production and reaction conditions are crucial in obtaining such size-dependent particle features. Particle size, chemical composition, crystallinity and shape can be controlled by temperature, pH-value, concentration, chemical composition, surface modifications and process control.

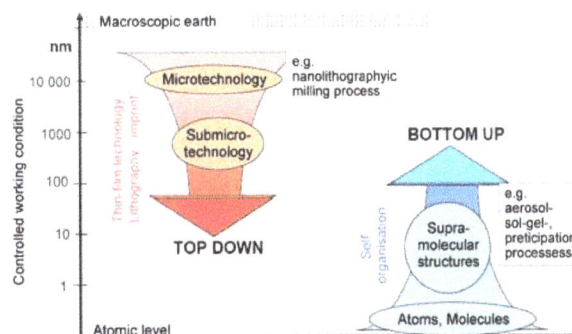

Figure: Methods of nanoparticle production: top-down and bottom-up.

Two basic strategies are used to produce nanoparticles: 'top-down' and 'bottom-up'. The term 'top-down' refers here to the mechanical crushing of source material using a milling process. In the 'bottom-up' strategy, structures are built up by chemical processes. The selection of the respective process depends on the chemical composition and the desired features specified for the nanoparticles.

## Top-Down/Mechanical-physical Production Processes

'Top-down' refers to mechanical-physical particle production processes based on principles of microsystem technology. The traditional mechanical-physical crushing methods for producing nanoparticles involve various milling techniques.

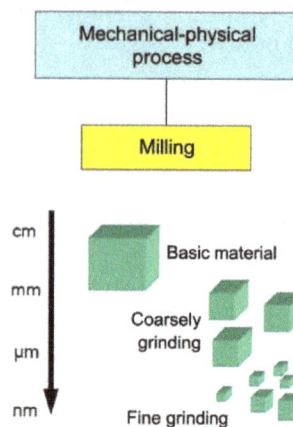

Figure: Overview of mechanical-physical nanoparticle production processes

## Milling Processes

The mechanical production approach uses milling to crush microparticles. This approach is applied in producing metallic and ceramic nanomaterials. For metallic nanoparticles, for example, traditional source materials (such as metal oxides) are pulverized using high-energy ball mills. Such mills are equipped with grinding media composed of wolfram carbide or steel.

Milling involves thermal stress and is energy intensive. Lengthier processing can potentially abrade the grinding media, contaminating the particles. Purely mechanical milling can be accompanied by reactive milling: here, a chemical or chemo-physical reaction accompanies the milling process.

Compared to the chemo-physical production processes, using mills to crush particles yields product powders with a relatively broad particle-size ranges. This method does not allow full control of particle shape.

## Bottom-up/Chemo-physical Production Processes

Bottom-up methods are based on physicochemical principles of molecular or atomic self-organization. This approach produces selected, more complex structures from atoms or molecules, better controlling sizes, shapes and size ranges. It includes aerosol processes, precipitation reactions and solgel processes.

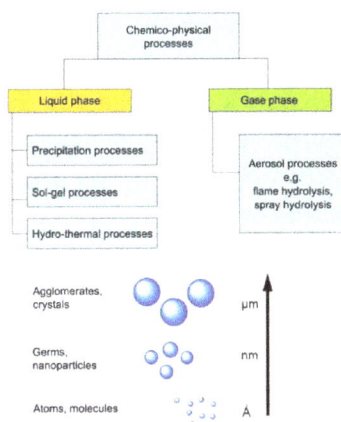

Figure: Chemo-physical processes in nanoparticle production

## Gas Phase Processes (Aerosol Processes)

Gas phase processes are among the most common industrial-scale technologies for producing nanomaterials in powder or film form.

Nanoparticles are created from the gas phase by producing a vapor of the product material using chemical or physical means. The production of the initial nanoparticles, which can be in a liquid or solid state, takes place via homogeneous nucleation.

Depending on the process, further particle growth involves condensation (transition from gaseous into liquid aggregate state), chemical reaction(s) on the particle surface and/or coagulation processes (adhesion of two or more particles), as well as coalescence processes (particle fusion). Examples include processes in flame-, plasma-, laser- and hot wall reactors, yielding products such as fullerenes and carbon nanotubes:

- In flame reactors, nanoparticles are formed by the decomposition of source molecules in the flame at relatively high temperatures (about 1200–2200 °C). Flame reactors are used today for the industrial-scale production of soot, pigment-titanium dioxide and silicon dioxide particles.

- In plasma reactors, plasma (ionized gas) provides the energy for the vaporization and for initializing the decomposition reactions.

- In laser reactors, lasers selectively heat the gaseous source material, utilizing its absorption wavelength, and decompose it to the desired product.

- In hot wall reactors, vaporization and condensation are applied. The source material is vaporized in an inert gas under low pressures (ca. 1 mbar). This removes the enriched gas phase from the hot zone. The particles created by the rapid cooling are collected on filters. Technically, hot wall reactors are used for example in producing nanoscale nickel- and iron powders.

- The chemical gas phase deposition process is used to directly deposit nanoparticles from the gas phase onto surfaces. Here, the source material is vaporized in a vacuum and condensed on a heated surface by a chemical reaction, i.e. deposited from the gas phase into the solid final product.

## Droplet Formation Containing Particles

Particles can also be produced from droplets using centrifugal forces, compressed air, sonic waves, ultrasound, vibrations, electrostatics and other methods. The droplets are transformed into a powder either through direct pyrolysis (thermal cleavage of chemical compounds) or via direct reactions with another gas. In spray pyrolysis, droplets of the source material are transported through a high-temperature field (flame, oven), which rapidly vaporizes the readily volatile components or leads to decomposition reactions. The formed particles are collected on filters.

## Liquid Phase Processes

The wet-chemical synthesis of nanomaterials typically takes place at lower temperatures than gas phase synthesis. The most important liquid phase processes in nanomaterial production are precipitation, sol-gel- processes and hydrothermal syntheses.

## Precipitation Processes

The precipitation of solids from a metal ion containing solution is one of the most frequently employed production processes for nanomaterials. Metal oxides as well as non-oxides or metallic nanoparticles can be produced by this approach. The process is based on reactions of salts in solvents. A precipitating agent is added to yield the desired particle precipitation, and the precipitate is filtered out and thermally post-treated.

In precipitation processes, particle size and size distribution, crystallinity and morphology (shape) are determined by reaction kinetics (reaction speed). The influencing factors include, beyond the concentration of the source material, the temperature, pH value of the solution, the sequence in which the source materials are added, and mixing processes.

A good size control can be achieved by using self-assembled membranes, which in turn serve as nanoreactors for particle production. Such nanoreactors include microemulsions, bubbles, micelles and liposomes. They are composed of a polar group and a non-polar hydrocarbon chain.

Micro-emulsions, for example, consist of two liquids that cannot be mixed with one another in the concentrations used, usually water and oil along with at least one tenside (substance that reduces the surface tension of liquids). In certain solvents this gives rise to small reactors in which nucleation and controlled particle growth take place. Particle size is determined by the size of the nanoreactors and, at the same time, particle agglomeration is prevented.

Micro-emulsion processes are often used to produce nanoparticles for pharmaceutical and cosmetics applications.

An additional process that is based on self-organized growth with templates and coatings is hydrothermal synthesis. Zeolites (microporous aluminum-silicon compounds) are produced from aqueous superheated solutions in autoclaves (airtight pressure chambers).

The partial vaporization of the solvent creates pressure in the autoclaves (several bars), triggering chemical reactions that differ from those under standard conditions, for example by altering the solubility. Nanoparticle formation and cavity shape can be controlled by adding templates. Templates are particles with bonds that enable the formation of certain forms and sizes.

# Sol-gel Processes

Sol-gel syntheses (production of a gel from powder-shaped materials) are wet-chemical processes for producing porous nanomaterials, ceramic nanostructured polymers as well as oxide nanoparticles. The synthesis takes place under relatively mild conditions and low temperatures.

The term sol refers to dispersions of solid particles in the 1-100 nm size range, which are finely distributed in water or organic solvents. In sol-gel processes, material production or deposition takes place from a liquid sol state, which is converted into a solid gel state via a sol-gel transformation. The sol-gel transformation involves a three-dimensional cross-linking of the nanoparticles in the solvent, whereby the gel takes on bulk properties. A controlled heat treatment in air can transform gels into a ceramic oxide material.

To start with, adding organic substances in the sol-gel process produces an organometallic compound from a solution containing an alcoxide (metallic compound of an alcohol, for example with silicon, titanium or aluminum). The pH value of the solution is adjusted with an acid or a base which, as a catalyst, also triggers the transformation of the alcoxide.

The subsequent reactions are hydrolysis (splitting of a chemical bond by water) followed by condensation and polymerization (reaction giving rise to many- or long-chained compounds from single-chained ones). The particles or the polymer oxide grow as the reaction continues, until a gel is formed. Due to the high porosity of the network, the particles typically have a large surface area, i.e. several hundred square meters per gram.

The course of hydrolysis and the polycondensation reaction depend on many factors: the composition of the initial solution, the type and amount of catalyst, temperature as well as the reactor- and mixing geometry.

For coatings, the alcoxide initial solution of the sol-gel process can be applied on surfaces of any geometry. After the wetting, the build-up of the porous network takes place through gel formation, yielding thicknesses of 50-500 nm. Thicker layers, suitable as membranes for example, are created by repeated wetting and drying. The sol-gel process can also be used to produce fibers. In all cases, gel formation is followed by a drying step. Figure 4 illustrates the different reaction and processing steps of the sol-gel process.

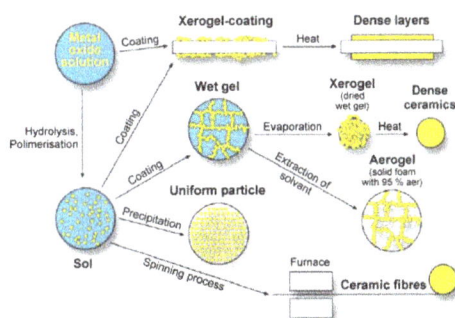

Figure: Reaction and processing steps in the sol-gel process.

A distinct advantage of the sol-gel process lies in the process ability of the sols and gels, depending on processing step, into powders, fibers, ceramics and coatings. Moreover, highly porous nanomaterials can be produced. Composites can be created by filling in these pores during or after gel production. The low process temperature also enables substances to be embedded into the gel during the synthesis step; these can then be stored or released in a controlled manner.

The disadvantage of the sol-gel process lies in the difficult-to-control synthesis and drying steps, which complicate scaling up the process. Moreover, organic contaminants can remain in the gel. The resulting necessary cleaning steps, drying and thermal post treatment makes this production process more complex than gas phase synthesis.

The disadvantage of the wet-chemical synthesis of nanomaterials is that the desired crystalline shapes often cannot be configured and that the thermal stability of the product powder is lower. This requires thermal post-treatment with repeated reduction of the particle surface. The advantage is that the liquid phase enables highly porous materials to be produced; this would normally not be possible in gas phase reactors due to the high temperatures.

With a few exceptions, gas phase processes also do not allow the production of organic nanoparticles. Liquid phase processes are particularly suited for the targeted production of monodisperse product powder (with uniform particle size).

## Magnetic Nanoparticle

Magnetic nanoparticles (MNP) have gained a lot of attention in biomedical and industrial applications due to their biocompatibility, easy of surface modification and magnetic properties. Magnetic nanoparticles can be utilized in versatile ways, very similar to those of nanoparticles in general. However, the magnetic properties of these particles add a new dimension where they can be manipulated upon application of an external magnetic field. This property opens up new applications where drugs that are attached to a magnetic particle to be targeted in the body using a magnetic field. Often, targeting is achieved by attaching a molecule that recognizes another molecule that is specific to the desired target area. This often requires a chemical recognition mechanism and does not succeed as designed. Therefore, magnetic nanoparticles can offer a solution to carry drugs to the desired areas in the body.

Magnetic nanoparticles, although may contain other elements, are often iron oxides. Most common iron oxides are magnetite ($Fe_3O_4$), maghemite ($\gamma$-$Fe_2O_3$), hematite ($\alpha$-$Fe_2O_3$) and geotite. Depending on the experimental conditions, one or more of the iron oxide phases may form. It is very important to carefully control the experimental conditions to ensure the presence of a single-phase.

Magnetic materials show a wide range of behaviors; at one end are the non-interacting spins in paramagnets and characterized by a temperature-dependent susceptibility $\left( v = M/H \propto 1/T \right)$ given by Curie's law. At the other end are the ferromagnets, with exchange interactions between spins, exhibiting hysteretic, M(H), behavior, and a finite coercivity, $H_C$ (M = 0), that is strongly dependent on the microstructure. Further, to minimize the overall magnetic energy, the material often forms domains, separated by domain walls with widths determined by the ratio of the exchange to anisotropy energies. However, if we reduce the size of any ferromagnet, we will ultimately reach a size where thermal energy ($k_BT$ ~ 25 meV, at 300 K) will compete with the prevailing anisotropy and randomize the magnetization direction such that for a typical measurement time (~100 s) the magnetization, M = 0, when no field is applied (H = 0). In other words, such materials show no coercivity ($H_C$ = 0), behaves similar to paramagnets but with very large moments, and are called super paramagnets. In practice, the randomization of the magnetization direction takes place by excitations over an energy barrier, $\varepsilon_B$ = KV, given by the product of the anisotropy constant, K, and the volume, V. As a first approximation, neglecting the applied field, the Arrhenius law can describe thermal excitations of the magnetic moment over an energy barrier, $\varepsilon_B$, with relaxation time, $\tau = \tau 0 \, exp \left( KV \, k_B^{-1} \, T^{-1} \right)$ Thus, super paramagnetic particles are defined by a characteristic diameter, Dsp, or a characteristic temperature called the blocking temperature, TB, such that, for a given measurement time, a sharp division from super paramagnetic to ferromagnetic behavior can be observed either as a function of size or temperature. Similar to a paramagnet, the magnetization response, M(H), of a super paramagnet is also given by the Langevin function. Note that because the relaxation time $\tau$, s, depends exponentially on the energy barrier, KV, to reproducibly control the magnetic behavior of super paramagnetic nanoparticles, especially under alternating fields, such as in MPI, tailored size and monodisperse size distributions in nanoparticle synthesis are required. Finally, for slightly larger particles, it is also important to consider the critical size that determines whether it is favorable to be uniformly magnetized (single domain), or to break into multiple domains to minimize their overall energy. Using simple models for domain stability in fine particles and bulk properties, one can determine the characteristic size, $D_{sd}$, up to which single domains are stable For particles with cubic anisotropy, the critical radius, $R_c = 9 / \mu_0 \, M_s^2$ with $D_{sd} = 2 R_c$, at which the nanoparticle breaks into multi-domains is a balance between the additional energy cost of introducing the domain wall and the reduction/gain in magnetostatic energy. This series of magnetic "phases" as a function of size for different ferromagnets and includes a "single domain" size ($D_{sd}$) below which the material will not support a multi-domain particle and a size ($D_{sp}$) defined by the super paramagnetic effect. Note that the characteristic size, $D_{sp}$, is determined by the measurement time (typically, 100 s is assumed); however, if the nanoparticles are subject to AC measurements, where the sampling time is inverse of the frequency, the observed $D_{sp}$ would be smaller than that shown in figure and inversely related to the sampling frequency.

## Synthesis

Several methods exist for preparing magnetic nanoparticle.

## Co-precipitation

Co-precipitation is a facile and convenient way to synthesize iron oxides (either $Fe_3O_4$ or $\gamma$-$Fe_2O_3$) from aqueous $Fe^{2+}/Fe^{3+}$ salt solutions by the addition of a base under inert atmosphere at room temperature or at elevated temperature. The size, shape, and composition of the magnetic nanoparticles very much depends on the type of salts used (e.g.chlorides, sulfates, nitrates), the $Fe^{2+}/Fe^{3+}$ ratio, the reaction temperature, the pH value and ionic strength of the media, and the mixing rate with the base solution used to provoke the precipitation. The co-precipitation approach has been used extensively to produce ferrite nanoparticles of controlled sizes and magnetic properties. A variety of experimental arrangements have been reported to facilitate continuous and large–scale co–precipitation of magnetic particles by rapid mixing. Recently, the growth rate of the magnetic nanoparticles was measured in real-time during the precipitation of magnetite nanoparticles by an integrated AC magnetic susceptometer within the mixing zone of the reactants.

## Thermal Decomposition

Magnetic nanocrystals with smaller size can essentially be synthesized through the thermal decomposition of alkaline organometallic compounds in high-boiling organic solvents containing stabilizing surfactants.

## Microemulsion

Using the microemulsion technique, metallic cobalt, cobalt/platinum alloys, and gold-coated cobalt/platinum nanoparticles have been synthesized in reverse micelles of cetyltrimethlyammonium bromide, using 1-butanol as the cosurfactant and octane as the oil phase.

## Flame Spray Synthesis

Using flame spray pyrolysis and varying the reaction conditions, oxides, metal or carbon coated nanoparticles are produced at a rate of > 30 g/h.

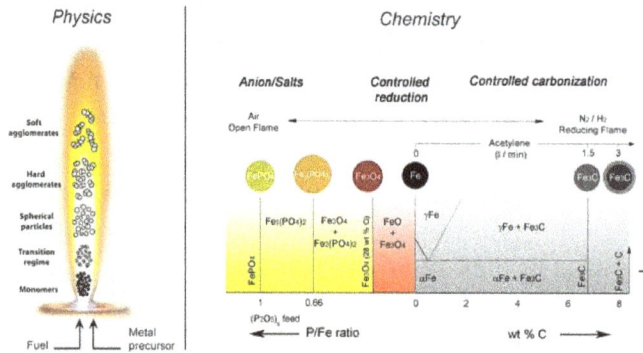

Various flame spray conditions and their impact on the resulting nanoparticles

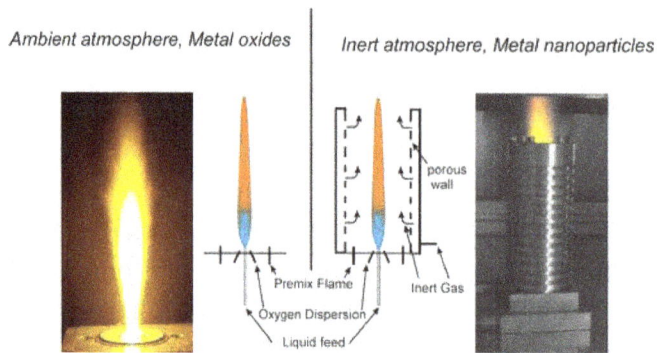

Operational layout differences between conventional and reducing flame spray synthesis

## Potential Applications

A wide variety of potential applications have been envisaged. Since magnetic nanoparticles are expensive to produce, there is interest in their recycling or for highly specialized applications. They are only used in scientific research. An industrial use has yet to be established.

The potential and versatility of magnetic chemistry arises from the fast and easy separation of the magnetic nanoparticles, eliminating tedious and costly separation processes usually applied in chemistry. Furthermore, the magnetic nanoparticles can be guided via a magnetic field to the desired location which could, for example, enable pinpoint precision in fighting cancer.

## Medical Diagnostics and Treatments

Magnetic nanoparticles have been examined for use in an experimental cancer treatment called magnetic hyperthermia in which an alternating magnetic field (AMF) is used to heat the nanoparticles. To achieve sufficient magnetic nanoparticle heating, the AMF typically has a frequency between 100–500 kHz, although significant research has been done at lower frequencies as well as frequencies as high as 10 MHz, with the amplitude of the field usually between 8-16kAm$^{-1}$.

Affinity ligands such as epidermal growth factor (EGF), folic acid, aptamers, lectins etc. can be attached to the magnetic nanoparticle surface with the use of various chemistries. This enables targeting of magnetic nanoparticles to specific tissues or cells. This strategy is used in cancer research to target and treat tumors in combination with magnetic hyperthermia or nanoparticle-delivered cancer drugs. Despite research efforts, however, the accumulation of nanoparticles inside of cancer tumors of all types is sub-optimal, even with affinity ligands. Willhelm et al. conducted a broad analysis of nanoparticle delivery to tumors and concluded that the median amount of injected dose reaching a solid tumor is only 0.7%. The challenge of accumulating large amounts of nanoparticles inside of tumors is arguably the biggest obstacle facing nanomedicine in general. While direct injection is used in some cases, intravenous injection is most often preferred to obtain a good distribution of particles throughout the tumor. Magnetic nanoparticles have a distinct advantage in that they can accumulate in desired regions via magnetically guided delivery, although this technique still needs further development to achieve optimal delivery to solid tumors.

Another potential treatment of cancer includes attaching magnetic nanoparticles to free-floating cancer cells, allowing them to be captured and carried out of the body. The treatment has been tested in the laboratory on mice and will be looked at in survival studies.

Magnetic nanoparticles can be used for the detection of cancer. Blood can be inserted onto a microfluidic chip with magnetic nanoparticles in it. These magnetic nanoparticles are trapped inside due to an externally applied magnetic field as the blood is free to flow through. The magnetic nanoparticles are coated with antibodies targeting cancer cells or proteins. The magnetic nanoparticles can be recovered and the attached cancer-associated molecules can be assayed to test for their existence.

Magnetic nanoparticles can be conjugated with carbohydrates and used for detection of bacteria. Iron oxide particles have been used for the detection of Gram negative bacteria like Escherichia coli and for detection of Gram positive bacteria like Streptococcus suis.

## Magnetic Immunoassay

Magnetic immunoassay (MIA) is a novel type of diagnostic immunoassay utilizing magnetic nanobeads as labels in lieu of conventional, enzymes, radioisotopes or fluorescent moieties. This assay involves the specific binding of an antibody to its antigen, where a magnetic label is conjugated to one element of the pair. The presence of magnetic nanobeads is then detected by a magnetic reader (magnetometer) which measures the magnetic field change induced by the beads. The signal measured by the magnetometer is proportional to the analyte (virus, toxin, bacteria, cardiac marker,etc.) quantity in the initial sample.

## Waste Water Treatment

Thanks to the easy separation by applying a magnetic field and the very large surface to volume ratio, magnetic nanoparticles have a potential for treatment of contaminated water.In this method, attachment of EDTA-like chelators to carbon coated metal nanomagnets results in a magnetic reagent for the rapid removal of heavy metals from solutions or contaminated water by three orders of magnitude to concentrations as low as micrograms per Litre. Magnetic nanobeads or nanoparticle clusters composed of FDA-approved oxide super paramagnetic nanoparticles (e.g. maghemite, magnetite) hold much potential for waste water treatment since they express excellent biocompatibility which concerning the environmental impacts of the material is an advantage compared to metallic nanoparticles.

## Supported Enzymes and Peptides

Enzymes, proteins, and other biologically and chemically active substances have been immobilized on magnetic nanoparticles. They are of interest as possible supports for solid phase synthesis.

This technology is potentially relevant to cellular labeling/cell separation, detoxification of biological fluids, tissue repair, drug delivery, magnetic resonance imaging, hyperthermia and magnetofection.

## Catalyst Support

Magnetic nanoparticles are of potential use as a catalyst or catalyst supports.In chemistry, a catalyst support is the material, usually a solid with a high surface area, to which a catalyst is affixed. The reactivity of heterogeneous catalysts occurs at the surface atoms. Consequently, great effort is made to maximize the surface area of a catalyst by distributing it over the support. The support may be inert or participate in the catalytic reactions. Typical supports include various kinds of carbon, alumina, and silica. Immobilizing the catalytic center on top of nanoparticles with a large surface to volume ratio addresses this problem. In the case of magnetic nanoparticles it adds the property of facile a separation. An early example involved a rhodium catalysis attached to magnetic nanoparticles .

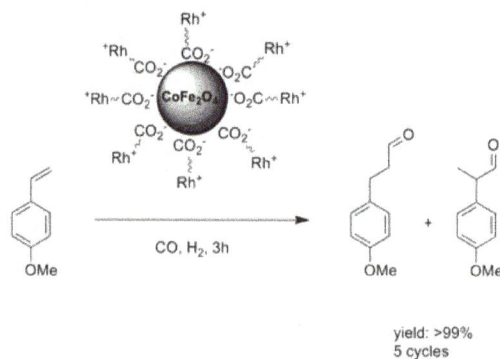

yield: >99%
5 cycles

In another example, the stable radical TEMPO was attached to the graphene-coated cobalt nanoparticles via a diazonium reaction. The resulting catalyst was then used for the chemoselective oxidation of primary and secondary alcohols.

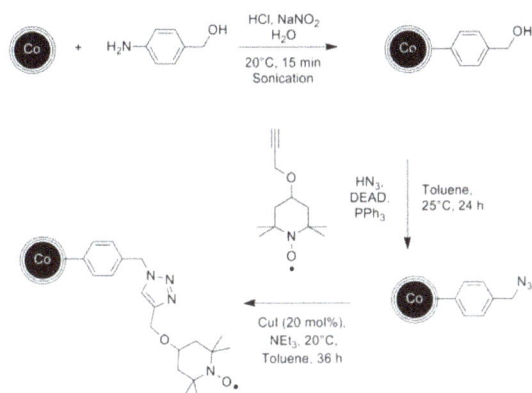

The catalytic reaction can be conducted in a continuous flow reactor instead of a batch reactor with no remains of the catalyst in the end product. Graphene coated cobalt nanoparticles have been used for that experiment since they exhibit a higher magnetization than Ferrite nanoparticles, which is essential for a fast and clean separation via external magnetic field.

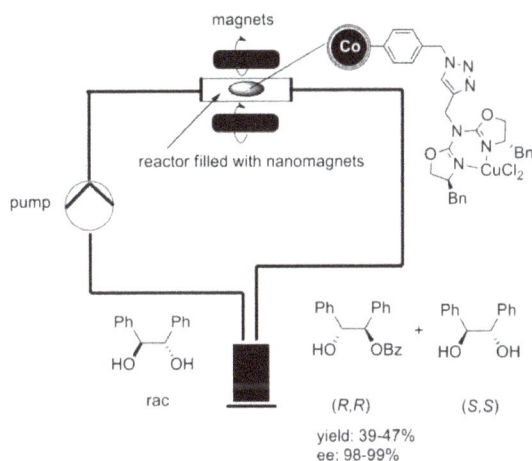

## Biomedical Imaging

There are many applications for iron-oxide based nanoparticles in concert with magnetic resonance imaging. Magnetic CoPt nanoparticles are being used as an MRI contrast agent for transplanted neural stem cell detection.

## Cancer Therapy

In magnetic fluid hyperthermia, nanoparticles of different types like Iron oxide, magnetite, maghemite or even gold are injected in tumor and then subjected under a high frequency magnetic field. These nanoparticles produce heat that typically increases tumor temperature to 40-46 °C, which can kill cancer cells.Another major potential of magnetic nanoparticles is the ability to combine heat (hyperthermia) and drug release for a cancer treatment. Numerous studies have shown particle constructs that can be loaded with a drug cargo and magnetic nanoparticles. The most prevalent construct is the "Magnetoliposome", which is a liposome with magnetic nanoparticles typically embedded in the lipid bilayer. Under an alternating magnetic field, the magnetic nanoparticles are heated, and this heat permeabilizes the membrane. This causes release of the loaded drug. This treatment option has a lot of potential as the combination of hyperthermia and drug release is likely to treat tumors better than either option alone, but it is still under development.

### Information Storage

A promising candidate for high-density storage is the face-centered tetragonal phase FePt alloy. Grain sizes can be as small as 3 nanometers. If it's possible to modify the MNPs at this small scale, the information density that can be achieved with this media could easily surpass 1 Terabyte per square inch.

### Genetic Engineering

Magnetic nanoparticles can be used for a variety of genetics applications. One application is the rapid isolation of mRNA. In one application, the magnetic bead is attached to a poly T tail. When mixed with mRNA, the poly A tail of the mRNA will attach to the bead's poly T tail and the isolation takes place simply by placing a magnet on the side of the tube and pouring out the liquid. Magnetic beads have also been used in plasmid assembly. Rapid genetic circuit construction has been achieved by the sequential addition of genes onto a growing genetic chain, using nanobeads as an anchor. This method has been shown to be much faster than previous methods, taking less than an hour to create functional multi-gene constructs in vitro.

### Physical Modeling

There are a variety of mathematical models to describe the dynamics of the rotations of magnetic nanoparticles. Simple models include the Langevin function and the Stoner-Wohlfarth model which describe the magnetization of a nanoparticle at equilibrium. The Debye/Rosenszweig model can be used for low amplitude or high frequency oscillations of particles, which assumes linear response of the magnetization to an oscillating magnetic field. Non-equilibrium approaches include the Langevin equation formalism and the Fokker-Planck equation formalism, and these have been developed extensively to model applications such as magnetic nanoparticle hyperthermia, magnetic nanoparticle imaging (MPI), magnetic spectroscopy and biosensing etc.

## Ferrite Nanoparticle

Ferrite nanoparticles with nominal composition $Me_{0.5}Fe_{2.5}O_4$ (Me = Co, Fe, Ni or Mn) have been successfully prepared by the wet chemical method. The obtained particles have a mean diameter of $11–16 \pm 2$ nm and were modified to improve their magnetic properties and chemical activity. The surface of the pristine nanoparticles was functionalized afterwards with $-COOH$ and $-NH_2$ groups to obtain a bioactive layer.

Super paramagnetism, collective magnetic excitations, low saturation magnetization and enhanced coercivity, metastable cation distributions, etc., are some of the phenomena which have been observed in nanoparticles of various ferrites. Additionally, when the particle size is controlled to lie within the single-domain regime, domain wall resonance is suppressed and the material can work at higher frequencies. Several physical and chemical methods, such as ball-milling (HEBM), inert gas condensations, co-precipitation, hydrothermal, micro-emulsion, process mechano-chemical synthesis and sol–gel auto-combustion, were reported to obtain ferrite nanoparticles. Table gives the summary with comparison of the synthesis methods.

Table: Comparison of the synthesis methods

| Synthesis method | Co-precipitation | Thermal decomposition | Micro-emulsion | Hydrothermal | Sol-gel |
|---|---|---|---|---|---|
| Synthesis | Very simple, ambient condition | Complicated, inert atmosphere | Complicated, ambient condition | Simple, high pressure | Simple, air atmosphere |
| React. tem[$°C$] | 20-90 | 100-320 | 20-50 | 220 | 60-80 |
| React. period | minutes | Hours/day | Hour | Hour's ca. days | Hours |
| Solvent | Water | Organic Compound | Organic Compound | Water-ethanol | Water |
| Surface capping agent | Needed, added during or after reaction | Needed, added during or after reaction | Needed, added during or after reaction | Needed, added during or after reaction | No Need |
| Size distribution | relatively narrow | Very narrow | Relatively narrow | Very narrow | Relatively narrow |
| Shape control | not good | Very good | Good | Very good | Very good |
| Yield | highly scalable | High/scalable | Low | Medium | High/scalable |

Among the several chemical processes, sol–gel, have acquired great importance during the last years, due to their potential for producing very pure and homogeneous nanostructures, with relatively large quantities of final product and low cost. It is probably the most effective method for the synthesis of homogeneous nanoparticles which can be prepared at relatively low temperatures. It consists of hydrolysis and condensation reactions of metal precursors leading to the formation of three-dimensional inorganic networks. Hydroxyl groups (M–OH) are formed during hydrolysis. These groups subsequently condense into strong and rigid metal-Oxo-metal bridges (M–O–M). These bridges are polymerized during heat treatment whereas they form a gel which is converted into dense particles.

In particular, the synthesis by combustion reaction technique has been shown to have great potential in the preparation of ferrites. This process is quite simple and involves an exothermic and self-sustaining chemical reaction between the metal salts and a suitable organic fuel, usually urea, glycine, hydrazides, dl-alanine, tartaric acid, citric acid.

The sol-gel combustion method presents some advantages compared to other methods: reagents are very simple compounds, special equipment is not required (brosil/pyrex containers are used), dopants can be easily introduced into the final product, and agglomeration of powders remains limited. This method uses the energy produced by the exothermic decomposition of a redox mixture of metal nitrates with an organic compound. In the combustion mixture nitrates and the organic compounds behave similarly to conventional oxidants and fuels.

The reaction is carried out by dissolving metal nitrates and fuel in a minimum amount of water in a borosil/pyrex dish and heating the mixture in order to evaporate water in excess. The resulting viscous liquid foams, ignites, and undergoes a self-sustained combustion, producing gases containing the oxide product. During the combustion, exothermic redox reactions associated with nitrate decomposition and fuel oxidation take place. Gases such as $N_2$ and $CO_2$ evolve, favoring the formation of fine particle ashes after only a few minutes. The properties of the final product (particle size, surface area, and porosity) depend on the way combustion is conducted. The departure of gases favors the desegregation of the products (increasing the porosity) and heat dissipation (inhibiting the sintering of the products). The exothermicity of the combustion is controlled by the nature of the fuel and the ratio oxidizer/fuel. Fuels are organic compounds, frequently hydrazine derivatives, with N-N bonds that undergo highly exothermic combustion. When fuels are heated in the presence of nitrates, the mixture of gases generated is always hypergolic: it reacts readily with evolvement of heat.

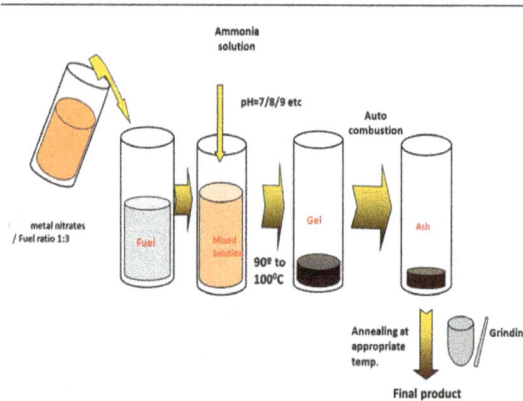

Figure: General Flow chart of sol-gel auto-combustion method

The stoichiometric composition of the metal nitrate-fuel mixtures is given by the equivalence ratio $\varphi e$, which reflects the relative ratio of intra-molecular fuel/oxidizer, considering the total reducing and oxidizing power of both fuel and oxidizer compounds. The mixture is considered fuel rich when $\varphi e > 1$, fuel an when 1 %, and stoichiometrically balanced when the $\varphi e < 1$. Other factors influencing combust ion are the evaporation time, the total mass of the redox mixture, the dish capacity, and the amount of water used to dissolve the reagents. Figure represents the general steps used in the sol-gel auto-combustion technique.

## Applications of Ferrite

Ferrite applications can be categorized in several different manners. First, they can be classified by market:

1) Consumer-entertainment.

2) Automotive.

3) Electrical Appliances.

4) Specialty and Custom- aircraft, microwave devices, recording heads.

5) Telecommunications- circuit components, power supplies.

The type of market aimed at will usually determine the cost of the magnetic material or component. The cost is lowest for the first category and successively higher for the remaining ones. Still another form of categorizing is by function and this maybe, for our purposes, the best way.

1. Voltage and current multipliers-Transformers.

2. Impedance Matching.

3. Inductor in LC circuit.

4. Filter to remove any unwanted frequencies- Wide band transformer, channel filter, EMI suppression.

5. Output choke- remove ac component from D.C.

6. Bi-stable element in a binary memory device- recording media.

7. Magnetic head- Write or read data on tape or disk.

8. Microwave devices.

9. Delay lines.

Another means of classification related to the application is made according to frequency;

1. D.C.-Permanent magnets and D.C. motors, generators and other D.C. devices.

2. Line frequencies-50-60 Hz.

3. Aircraft frequencies-400 Hz.

4. Audio Frequencies-to 20,000 Hz.

5. High frequency power- 25000-100000 Hz. and climbing.

6. High frequency telecommunications-100,000 Hz. to 100 MHz.

7. Microwave and Radar-1 GHz.

In general, the frequency used is also an indication of the size of the component. The lower the frequency, the larger the size while conversely, high frequency components tend to be smaller. Recently, nanoparticle ferrites with a high surface to volume ratio have received much attention due to their useful, electrical and magnetic properties used in magnetic fluid, information storage and medical diagnostics like targeted drug delivery, magnetic resonance imaging (MRI), hyperthermia etc. figure represents the graphical representation of several synthesis methods, properties and applications of nickel ferrite nanoparticles.

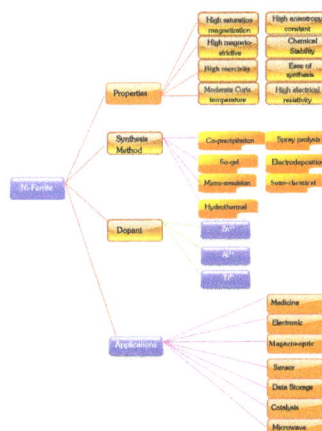

Figure: Properties, synthesis methods and applications of nickel ferrite nanoparticles

# References

- Reeves, Daniel B. (2017). "Nonlinear Nonequilibrium Simulations of Magnetic Nanoparticles". Magnetic Characterization Techniques for Nanomaterials. Springer, Berlin, Heidelberg. pp. 121–156. doi:10.1007/978-3-662-52780-1_4. ISBN 978-3-662-52779-5

- Nanoparticle, science: britannica.com, Retrieved 09 July 2018

- Reeves, Daniel B.; Weaver, John B. (2014). "Approaches for Modeling Magnetic Nanoparticle Dynamics". Critical Reviews™ in Biomedical Engineering. 42 (1): 85–93. arXiv:1505.02450. doi:10.1615/CritRevBiomedEng.2014010845. ISSN 0278-940X. PMC 4183932

- How-nanoparticles-are-made: nanowerk.com, Retrieved 28 May 2018

- K.Norén, Katarina; M. Kempe (2009). "Multilayered Magnetic Nanoparticles as a Support in Solid-Phase Peptide Synthesis". International Journal of Peptide Research and Therapeutics. 15 (4): 287–292. doi:10.1007/s10989-009-9190-3

- Estelrich, Joan; et al. (2015). "Iron Oxide Nanoparticles for Magnetically-Guided and Magnetically-Responsive Drug Delivery". Int. J. Mol. Sci. 16 (12): 8070–8101. doi:10.3390/ijms16048070. PMC 4425068. PMID 25867479

# 3

# Structure of Nanomaterials

An understanding of nanomaterials requires an in-depth study of nanomaterial structure. The significant aspects of the structure of nanomaterials such as quantum dot, nanowire, nanorod, nanocrystalline materials, etc. have been extensively discussed in this chapter.

## Nanomaterials Structure

The microstructure is defined by the type, crystal structure, number, shape and topological arrangement of phases and defects such as point defects, dislocations, stacking faults or grain boundaries in a crystalline material. Nanocrystalline materials are polycrystalline solids with grain sizes below 100 nm. Grains as well as pores, interfaces and other defects are of similar dimensions. This nano-specific microstructure (nanostructure) leads to chemical and physical size effects. It is a prerequisite for the understanding of properties of nanomaterials to have a detailed knowledge of the structure from the atomic/molecular (local) to the crystal structure (long range order) and to the microstructure (mesoscopic scale and defect structure). Consequently, various analytical techniques are required to characterize the nanomaterials on all length scales.

### Particle Size and Morphology

Scattering and imaging methods such as line broadening in X-ray diffraction, small angle X-ray and neutron scattering and transmission electron microscopy, are employed to determine average particle and grain size, their distributions and the morphology of the particles. Additionally, nitrogen adsorption, light scattering, atomic probe techniques and mass spectroscopy provide complementary information on these size related parameters. The method of choice depends on the type of material (powder, solid and liquid dispersion, consolidated and sintered ceramic, etc.). Care should be taken in comparing the average sizes determined by these methods because different size related parameters are determined (e.g. column length in XRD line broadening) and the average values are based on different weight functions.

### Local Structure

Information on the local structure can be provided by spectroscopic methods such as NMR or EXAFS spectroscopy but is also contained in diffuse scattering. Nanocrystalline materials are heterogeneously disordered with a large fraction of atoms located in surfaces and interfaces. In case of pure, tetragonal nanocrystalline zirconia with a particle size of 5 nm it was shown that atoms located at particle surfaces have a higher degree of disorder compared to atoms in the core of the particles due to more degrees of freedom. Primary processes during the particle formation by CVS and the evolution of the microstructure during sintering occur on the local, molecular level. Segregation of aluminium atoms was identified as the mechanism for the grain growth inhibition during sintering of zirconia doped with alumina by Reverse Monte Carlo analysis of EXAFS spectra. Similarly, an inhomogeneous distribution of yttrium in zirconia was found and correlated to a low sinterability.

## Crystal Structure

X-ray, electron and neutron diffraction are extensively used to characterize the crystal structure which has been observed to be size dependent in many cases. The line broadening at very small crystallite sizes limits the accuracy of crystal structure determination, i.e. difficulty in distinguishing between tetragonal and cubic zirconia. In general it is observed for CVS nanopowders that all diffraction reflexes are present and that the background is very low. This indicates a high degree of crystallinity and low defect density within the individual particles. An exemption is nanocrystalline silicon carbide which exhibits stacking faults and twinning, especially when synthesized at low temperature. From Rietveld refinement not only the phase composition, lattice parameters and positions of atoms in the unit cell, but also crystallite size and the microstrain can be extracted.

Depending on external conditions such as temperature or pressure, atoms may arrange themselves in various ways in lattice structures. Therefore, for some materials with the same proportions of contained elements, different crystal structures exist. Nanomaterials with different crystal structures may differ in important physicochemical properties (e.g. reactivity or photocatalytic activity). Accordingly, in such cases only a specific crystal structure is used for a given application. The different crystal structures are not only relevant for the various technical applications, but can also affect the behavior and toxicity of nanomaterials.

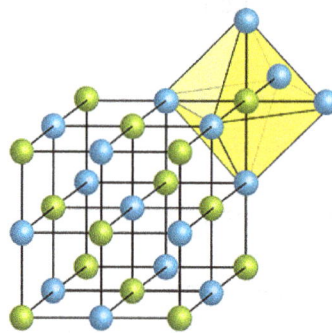

An example for such a material is titanium dioxide ($TiO_2$). To sun screen or as white pigment in paints it is mainly added in the rutile form, whereas for photocatalytic applications (such as self-cleaning surfaces) predominantly the anatase form is used.

Different crystal structures of titanium dioxide

Anatase          Rutile          Brookite

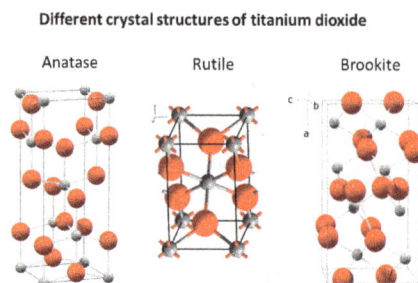

In addition to titanium dioxide a number of other nanomaterials come in different crystal structures, e.g. iron oxide, silicon dioxide or carbon. The latter can occur in the form of diamond or graphite (a stack of multiple graphene layers). Because of its hardness, diamond is used for example for cutting concrete, whereas graphite is used a lubricant due to the good deformability of its crystal lattice. Consequently, the varying material properties of different crystal structures open a broad spectrum of applications.

The various crystal structures of nanomaterials have to be taken into account when analyzing potential adverse effects on organisms or the environmental behavior. The physico-chemical properties of the materials are known to differ depending on crystal structure.

As was demonstrated for titanium dioxide, the anatase and rutile modifications differ in their toxic effects on organisms and cells. Anatase shows a toxic effect on cells and organisms, while rutile was found to be non-toxic. On the one hand, the photocatalytic effect of the anatase titanium dioxide, which contributes to the formation of reactive oxygen species (ROS) is discussed as potential reason. On the other hand, it is known that the rutile form is poorly soluble in water. Therefore, it is believed that the rutile titanium dioxide may agglomerate and subsequently sediment in aqueous test media, and hence organisms and cells are exposed to particles only for short periods. Likewise, such influences of the crystal structure on the toxic effects have been described for other nanomaterials.

Additionally, the different crystal structures can influence other nanomaterial properties such as solubility or sorption of other substances.

## Electronic Structure of Nanomaterials

The electronic and chemical structure of carbon materials are highly dependent on the hybridization of carbon atoms, but also on their functionalization and their nanostructure. We are also very interested in the effect of solvation on their electronic and chemical properties.

Several classes of materials are investigated in our group:

## Diamond-based Materials

Diamond is a wide bandgap semiconductor (5.5 eV), which has exceptional electronic properties, such as a negative electron affinity when its surface is hydrogenated. It can also be doped by heteroatoms such as boron, nitrogen or phosphorus. Nevertheless, the impact of nanostructuration (nanoparticles, nanoporous diamond) and doping on the diamond electronic properties are not fully understood yet. We are probing different kind of diamond materials using X-ray absorption and emission spectroscopies at the carbon K edge to probe unoccupied and occupied electronic states, respectively, and gain new insights into the link between structure and electronic properties.

We are also probing nanodiamonds directly in solution using microjet and flow cell systems to observe the effect of solvation on their electronic structure. In particular, we have found that graphitic surface states are reacting with water molecules, inducing the formation of holes in the diamond valence band.

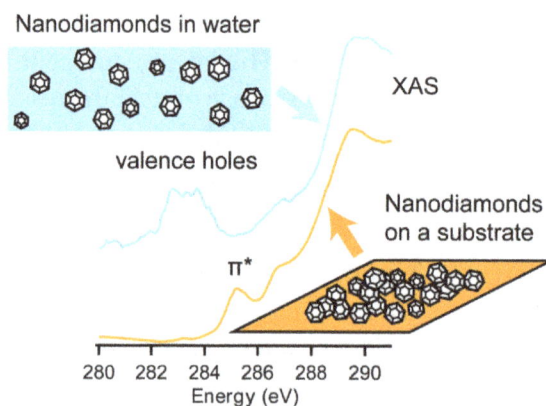

Figure: Carbon K edge XAS measurements of dry and dispersed detonation nanodiamonds.

## Carbon Dots

Many different carbon dots and graphene quantum dots, having mainly amorphous or graphitic hybridization have been reported in the literature.

## Carbon-containing 2D Materials

In addition to fully carbon-based nanomaterials, we are also interested in the characterization of hybrid 2D nanomaterials. We are investigating polymeric carbon nitride ($C_3N_4$), which is a promising photocatalyst for solar fuel production. We are also investigating electronic properties of titanium carbide ($Ti_3C_2$) which has been proposed as an outstanding material for electrochemical storage.

# Nanowire

Nanowire is a solid rod-like material or structure with diameter on the order of nanometers. Similar to conventional wires, they are also manufactured from semiconducting metal oxides, metals or carbon nanotubes. Due to their size they exhibit unique thermal, chemical, electronic, optical and mechanical properties which are not found in bulk materials and which have their related fields of applications.

Nanowire is produced under controlled conditions, and can be manufactured via several processes, such as vapor deposition, vapor-liquid synthesis and suspension. Nanowire can be metallic, insulating or semiconducting.

Nanowire exhibits high flexibility and is high in strength and uniform morphology. Metallic nanowire exhibits enhanced magnetic coercivity compared to their bulk counterparts. As for thermoelectric properties, metallic nanowire shows high Seeback coefficient because of the enhanced density of electronic states. Thus it can conduct heat or electricity substantially higher than any bulk material. When it comes to electrical properties, the crystalline structure of nanowire increases the electrical properties by many fold. The large surface area of nanoparticles provides motivating catalytic properties for nanowire. Considering optical properties, metallic nanowire shows unique plasmon absorption effect. On-linear properties are also shown by nanowire arrays.

The unique capabilities and features of nanowire hold lot of promise for applications in different fields like optics, electronics and magnetism. The high aspect ratio, high number of surface atoms and enhanced surface-to-volume ratio makes nanowire attractive for sensor applications such as nanosensors. They are also used in small electronic circuits, transistors, memory devices and quantum instruments. Nanowire is also used in producing nanoprobes and nanophotons.

## Properties of Nanowire

## Mechanical Property

The enormous amount of grain boundaries in a bulk material are made of nanoparticles which allow extending the grain boundaries sliding leads to high flexibility. The figure below consists of gate insulator device and substrate which involve in the operation of the mechanical property of nanowire.

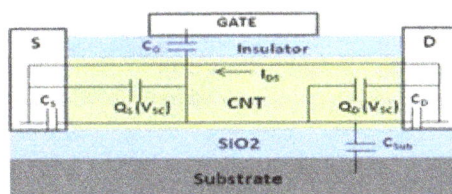

## Magnetic Property

In the magnetic property of nanoparticles the energy of magnetic anisotropy might be that miniature that the vector of magnetization fluctuates thermally, this is called super magnetism. Such materials are free from reminisce and coercitivity. Touching super magnetic particles are losing this special property by get in touch with expect the particles are kept at distance. Unusually electronic and magnetic characteristics are established at non zero temperature such as the metal insulator alteration in metal oxides non Fermi liquid performance of highly interrelated f-electron compound, uncharacteristic symmetry state of high-Tc superconductor device. Combining the particles with high energy of anisotropy with a super magnetic may over leads to a new class of permanent magnetic materials.

## Catalytic Property

Due to large surface of area, the nanoparticles which are made of transition materials oxide exhibits motivating catalytic properties. In some of the special cases catalysis may be improved and additional specific by decorating these particles with gold and platinum duster.

## Optical Property

 In optical property the allotment of non-agglomerated nanoparticles in a polymer are used to the directory of refraction. In addition, such a procedure may manufacture material with non-linear optical properties or visual property. Gold and Cd se nanoparticles in glass leads to red or orange coloration semi conducting nanopractices and some oxide polymer nanocompositor exhibits fluorescence performance is blue shift with decreasing particles size. Faraday rotation is one of the magneto optical effects extremely premeditated for Ferro fluid.

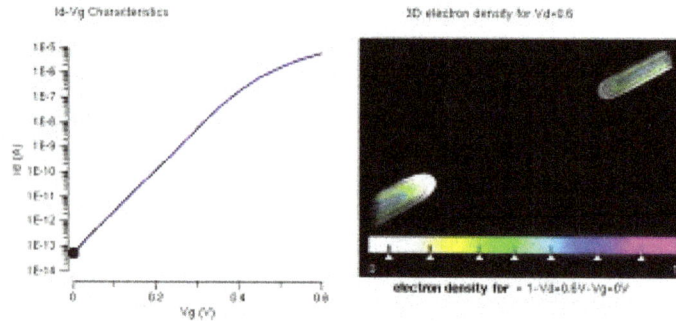

## Synthesis

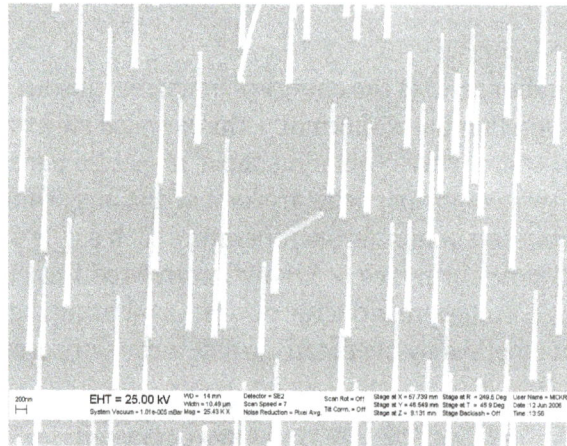

An SEM image of epitaxial nanowire heterostructures grown from catalytic gold nanoparticles.

There are two basic approaches to synthesizing nanowires: top-down and bottom-up. A top-down approach reduces a large piece of material to small pieces, by various means such as lithography, milling or thermal oxidation. A bottom-up approach synthesizes the nanowire by combining constituent adatoms. Most synthesis techniques use a bottom-up approach. Initial synthesis via either method may often be followed by a nanowire thermal treatment step, often involving a form of self-limiting oxidation, to fine tune the size and aspect ratio of the structures.

Nanowire production uses several common laboratory techniques, including suspension, electrochemical deposition, vapor deposition, and VLS growth. Ion track technology enables growing homogeneous and segmented nanowires down to 8 nm diameter. As nanowire oxidation rate is controlled by diameter, thermal oxidation steps are often applied to tune their morphology.

## Suspension

A suspended nanowire is a wire produced in a high-vacuum chamber held at the longitudinal extremities. Suspended nanowires can be produced by:

- The chemical etching of a larger wire.

- The bombardment of a larger wire, typically with highly energetic ions.

- Indenting the tip of a STM in the surface of a metal near its melting point, and then retracting it.

## Vls Growth

A common technique for creating a nanowire is vapor-liquid-solid method (VLS), which was first reported by Wagner and Ellis in 1964 for silicon whiskers with diameters ranging from 100s of nm to 100s of μm. This process can produce high-quality crystalline nanowires of many semiconductor materials, for example, VLS–grown single crystalline silicon nanowires (SiNWs) with smooth surfaces could have excellent properties, such as ultra-large elasticity. This method uses a source material from either laser ablated particles or a feed gas such as silane.

VLS synthesis requires a catalyst. For nanowires, the best catalysts are liquid metal (such as gold) nanoclusters, which can either be self-assembled from a thin film by dewetting, or purchased in colloidal form and deposited on a substrate.

The source enters these nanoclusters and begins to saturate them. On reaching supersaturation, the source solidifies and grows outward from the nanocluster. Simply turning off the source can adjust the final length of the nanowire. Switching sources while still in the growth phase can create compound nanowires with super-lattices of alternating materials.

A single-step vapour phase reaction at elevated temperature synthesises inorganic nanowires such as $Mo_6S_{9-x}I_x$. From another point of view, such nanowires are cluster polymers.

## Solution-phase Synthesis

Solution-phase synthesis refers to techniques that grow nanowires in solution. They can produce nanowires of many types of materials. Solution-phase synthesis has the advantage that it can produce very large quantities, compared to other methods. In one technique, the polyol synthesis, ethylene glycol is both solvent and reducing agent. This technique is particularly versatile at producing nanowires of gold, lead, platinum, and silver.

The supercritical fluid-liquid-solid growth method can be used to synthesize semiconductor nanowires, e.g., Si and Ge. By using metal nanocrystals as seeds, Si and Ge organometallic precursors are fed into a reactor filled with a supercritical organic solvent, such as toluene. Thermolysis results in degradation of the precursor, allowing release of Si or Ge, and dissolution into the metal nanocrystals. As more of the semiconductor solute is added from the supercritical phase (due to a concentration gradient), a solid crystallite precipitates, and a nanowire grows uniaxially from the nanocrystal seed.

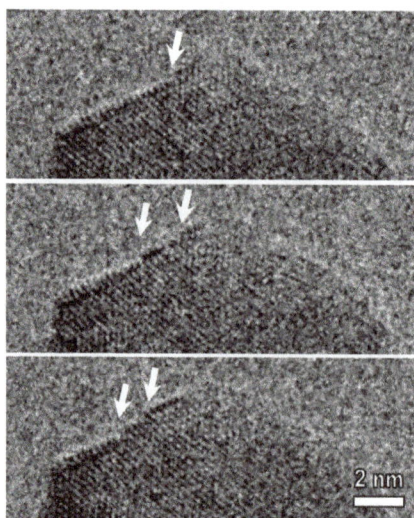

In situ observation of CuO nanowire growth

## Non-catalytic Growth

Nanowires can be also grown without the help of catalysts, which gives an advantage of pure nanowires and minimizes the number of technological steps. The simplest methods to obtain metal oxide nanowires use ordinary heating of the metals, e.g. metal wire heated with battery, by Joule heating in air can be easily done at home. The vast majority of nanowire-formation mechanisms are explained through the use of catalytic nanoparticles, which drive the nanowire growth and are either added intentionally or generated during the growth. However the mechanisms for catalyst-free growth of nanowires (or whiskers) were known from 1950s. Spontaneous nanowire formation by non-catalytic methods were explained by the dislocation present in specific directions or the growth anisotropy of various crystal faces. More recently, after microscopy advancement, the nanowire growth driven by screw dislocations or twin boundaries were demonstrated. The picture on the right shows a single atomic layer growth on the tip of CuO nanowire, observed by in situ TEM microscopy during the non-catalytic synthesis of nanowire.

## DNA-templated Metallic Nanowire Synthesis

An emerging field is to use DNA strands as scaffolds for metallic nanowire synthesis. This method is investigated both for the synthesis of metallic nanowires in electronic components and for bio-sensing applications, in which they allow the transduction of a DNA strand into a metallic nanowire that can be electrically detected. Typically, ssDNA strands are stretched, where after they are decorated with metallic nanoparticles that have been functionalized with short complementary ssDNA strands.

## Applications

## Electronic Devices

Nanowires can be used for transistors. Transistors are used widely as fundamental building element in today's electronic circuits. As predicted by Moore's law, the dimension of transistors is shrinking smaller and smaller into nanoscale. One of the key challenges of building future nanoscale transistors is ensuring good gate control over the channel. Due to the high aspect ratio, if the gate dielectric is wrapped around the nanowire channel, we can get good control of channel electrostatic potential, thereby turning the transistor on and off efficiently.

Due to the unique one-dimensional structure with remarkable optical properties, the nanowire also opens new opportunities for realizing high efficiency photovoltaic devices. Compared with its bulk counterparts, the nanowire solar cells are less sensitive to impurities due to bulk recombination, and thus silicon wafers with lower purity can be used to achieve acceptable efficiency, leading to the a reduction on material consumption.

To create active electronic elements, the first key step was to chemically dope a semiconductor nanowire. This has already been done to individual nanowires to create p-type and n-type semiconductors.

The next step was to find a way to create a p–n junction, one of the simplest electronic devices. This was achieved in two ways. The first way was to physically cross a p-type wire over an n-type wire. The second method involved chemically doping a single wire with different dopants along the length. This method created a p-n junction with only one wire.

After p-n junctions were built with nanowires, the next logical step was to build logic gates. By connecting several p-n junctions together, researchers have been able to create the basis of all logic circuits: the AND, OR, and NOT gates have all been built from semiconductor nanowire crossings.

In August 2012, researchers reported constructing the first NAND gate from undoped silicon nanowires. This avoids the problem of how to achieve precision doping of complementary nanocircuits, which is unsolved. They were able to control the Schottky barrier to achieve low-resistance contacts by placing a silicide layer in the metal-silicon interface.

It is possible that semiconductor nanowire crossings will be important to the future of digital computing. Though there are other uses for nanowires beyond these, the only ones that actually take advantage of physics in the nanometer regime are electronic.

In addition, nanowires are also being studied for use as photon ballistic waveguides as interconnects in quantum dot/quantum effect well photon logic arrays. Photons travel inside the tube, electrons travel on the outside shell.

When two nanowires acting as photon waveguides cross each other the juncture acts as a quantum dot.

Conducting nanowires offer the possibility of connecting molecular-scale entities in a molecular computer. Dispersions of conducting nanowires in different polymers are being investigated for use as transparent electrodes for flexible flat-screen displays.

Because of their high Young's moduli, their use in mechanically enhancing composites is being investigated. Because nanowires appear in bundles, they may be used as tribological additives to improve friction characteristics and reliability of electronic transducers and actuators.

Because of their high aspect ratio, nanowires are also uniquely suited to dielectrophoretic manipulation, which offers a low-cost, bottom-up approach to integrating suspended dielectric metal oxide nanowires in electronic devices such as UV, water vapor, and ethanol sensors.

# Nanowire Lasers

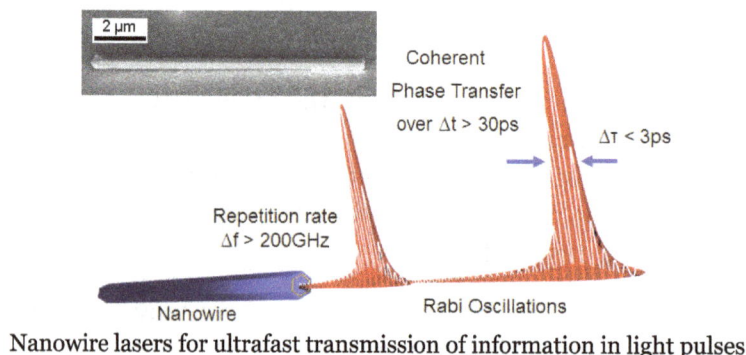

Nanowire lasers for ultrafast transmission of information in light pulses

Nanowire lasers are nano-scaled lasers with potential as optical interconnects and optical data communication on chip. Nanowire lasers are built from III–V semiconductor heterostructures, the high refractive index allows for low optical loss in the nanowire core. Nanowire lasers are subwavelength lasers of only a few hundred nanometers. Nanowire lasers are Fabry–Perot resonator cavities defined by the end facets of the wire with high-reflectivity, recent developments have demonstrated repetition rates greater than 200 GHz offering possibilities for optical chip level communications.

## Sensing of Proteins and Chemicals using Semiconductor Nanowires

In an analogous way to FET devices in which the modulation of conductance (flow of electrons/holes) in the semiconductor, between the input (source) and the output (drain) terminals, is controlled by electrostatic potential variation (gate-electrode) of the charge carriers in the device conduction channel, the methodology of a Bio/Chem-FET is based on the detection of the local change in charge density, or so-called "field effect", that characterizes the recognition event between a target molecule and the surface receptor.

This change in the surface potential influences the Chem-FET device exactly as a 'gate' voltage does, leading to a detectable and measurable change in the device conduction. When these devices are fabricated using semiconductor nanowires as the transistor element the binding of a chemical or biological species to the surface of the sensor can lead to the depletion or accumulation of charge carriers in the "bulk" of the nanometer diameter nanowire i.e. (small cross section available for conduction channels). Moreover, the wire, which serves as a tunable conducting channel, is in close contact with the sensing environment of the target, leading to a short response time, along with orders of magnitude increase in the sensitivity of the device as a result of the huge S/V ratio of the nanowires.

While several inorganic semiconducting materials such as Si, Ge, and metal oxides (e.g. $In_2O_3$, $SnO_2$, ZnO, etc.) have been used for the preparation of nanowires, Si is usually the material of choice when fabricating nanowire FET-based chemo/biosensors.

Several examples of the use of silicon nanowire(SiNW) sensing devices include the ultra-sensitive, real-time sensing of biomarker proteins for cancer, detection of single virus particles, and the detection of nitro-aromatic explosive materials such as 2,4,6 Tri-nitrotoluene (TNT) in sensitives superior to these of canines.Silicon nanowires could also be used in their twisted form, as electromechanical devices, to measure intermolecular forces with great precision.

## Limitations of Sensing with Silicon Nanowire FET Devices

Generally, the charges on dissolved molecules and macromolecules are screened by dissolved counterions, since in most cases molecules bound to the devices are separated from the sensor surface by approximately 2–12 nm (the size of the receptor proteins or DNA linkers bound to the sensor surface). As a result of the screening, the electrostatic potential that arises from charges on the analyte molecule decays exponentially toward zero with distance. Thus, for optimal sensing, the Debye length must be carefully selected for nanowire FET measurements. One approach of overcoming this limitation employs fragmentation of the antibody-capturing units and control over surface receptor density, allowing more intimate binding to the nanowire of the target protein. This approach proved useful for dramatically enhancing the sensitivity of cardiac biomarkers (e.g. Troponin) detection directly from serum for the diagnosis of acute myocardial infarction.

## Nanorod

Nanorods are one dimensional structure which provides a directed path for electrical transport and are used to control the band gap by varying the radius of rods and using the quantum size effect. The efficiency of quantum dot conjugated polymer Solar cell can be enhanced by using quantum confinement effect which will affect the length and width of nanorods leading to thinner devices for optimal absorption of incident light.

Gold nanorods are considered excellent candidates for biological sensing applications because the absorbance band changes with the refractive index of local material , allowing for extremely accurate sensing. In addition, nanorods with near-infrared absorption peaks can be excited by a laser at the absorbance band wavelength to produce heat, potentially allowing for the selective thermal destruction of cancerous tissues.

Nanoscale materials such as fullerenes, quantum dots and metallic nanoparticles have unique properties, because of their high surface area to volume ratio. Gold nanospheres and nanorods also have unique optical properties, because of the quantum size effect. Gold nanorods are cylindrical rods which range from less than ten to over forty nanometers in width and up to several hundred nanometers in length. These particles are typically characterized by their aspect ratio (length divided by width).

In order to study and exploit the unique properties of nanorods, it is necessary to have a robust extinction coefficient which can predict the concentration of a solution at a particular absorbance. It is difficult to accurately obtain a measure of nanoparticle (as opposed to metal atom) concentration in moles per liter. No spectroscopic device can provide concentration data, and only approximations are currently available.

Biomedical applications of nanoparticles require nanorods to be capped with biological molecules such as antibodies.

The El-Sayed method of nanorod concentration determination is currently the standard way of measuring extinction coefficients ($\varepsilon$), and involves the coupling of bulk gold concentration, Transmission Electron Microscopy (TEM) size analysis and absorbance data.

Synthesis of gold nanorods has recently undergone dramatic improvements. It is possible to produce high yields of nearly monodispersed short gold nanorods. First, a "seed" solution of spherical gold nanoparticles was prepared by adding the following:

| 0.01M $HAuCl_4.3H_2O$ | 0.250 |
|---|---|
| 0.1 M CTAB (cetyltrimethylammonium bromide) | 7.5 |
| 0.01 M $NaBH_4$ | 0.6 |

Table: Preparation of Gold seed

$NaBH_4$ reduces the gold salt to form nanoparticles, and cetyltrimethylammonium bromide (CTAB) is a surfactant which stabilizes the seeds to prevent aggregation. To make nanorods, the reagents in table are added in order from top to bottom. Gold rods are synthesized with a small amount of silver to control rod size and make short rods. CTAB is a directing surfactant; without it, only spheres would form. CTAB forms a rod shaped template that is filled with gold atoms as they are reduced by ascorbic acid. Ascorbic acid is a weaker reducing agent than $NaBH_4$, but in the presence of seeds and a CTAB template, it reduces gold ions at the seed surface.

## Synthesis of Nanorods

## General Idea is the same as the Growth of Nanorods (Seed-mediated Method)

To make surfactant-coated nanorods, the well-documented seed-mediated procedure developed by was employed. This method yields nanorods that are stabilized as verified from transmission electron microscopy (TEM) analysis.

## Slightly Change the Conditions when Growing Nanorods (Concentration of Different Reactants)

The nanorods were coated with a very thin (ca. 3-5 nm) silica film that:

i.   Improved the colloidal stability of the nanorods by reducing aggregation,

ii.  Improved the shape stability of the nanorods, and

iii. Allowed for further modification of the nanorod surface.

This silication method was first developed for citrate-stabilized gold NPs and has been applied successfully to gold nanorods.

## Cubes, Hexagon, Triangle, Tetropods and Branched

A two-step growth method has been developed to grow nanorods by changing the oxygen content in gas mixture during nucleation and growth steps. This is based on our systematic studies of nucleation and growth under different conditions. Due to the large lattice mismatch (~;18%) between molecule and sapphire, the nucleation of molecule on sapphire follows the three-dimensional island growth; that is, the Volmer–Weber mode, as reported by Yamauchi et al. in their observation of plasma-assisted epitaxial growth of molecule on sapphire. At high temperature, the nucleation of nanorods islands on the surface of substrates depends strongly on the amount of active oxygen. When grown entirely in 90% oxygen plasma, nanorods have a high nucleation density and forms.

Direct chemical synthesis and a combination of ligands are all that are required for production and shape control of the nanorods. Ligands also bond to different facets of the nanorod with varying strengths. This is how different faces of nanorods can be made to grow at different rates, thereby producing an elongated object of a certain desired shape.

Step 1: Symmetry breaking in FCC metals

Step 2: Preferential surfactant binding to specific creystals faces

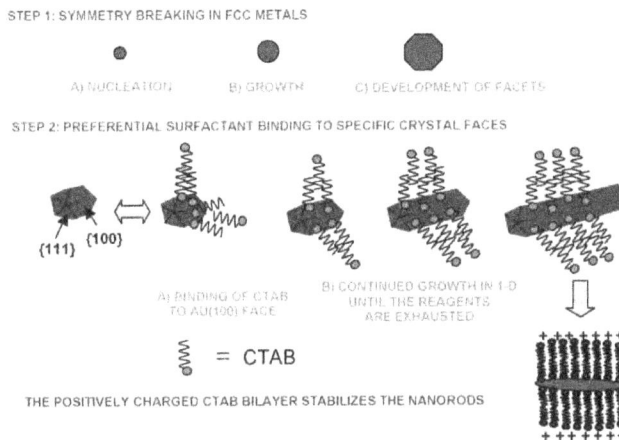

We synthesized silver nanorods with the average length of 280 nm and diameters of around 25 nm were synthesized by a simple reduction process of silver nitrate in the presence of polyvinyl alcohol (PVA) and investigated by means of scanning electron microscopy (SEM), X-ray diffraction (XRD), transmission-electron microscopy (TEM).

It was found out that both temperature and reaction time are important factors in determining the morphology and aspect ratios of nanorods. TEM images showed the prepared silver nanorods have a face centered shape (fcc) with fivefold symmetry consisting of multiply twinned face centered cubes as revealed in the cross-section observations. The five-fold axis, i.e. the growth direction, normally goes along the (111) zone axis direction of the basic fcc Ag-structure. Preferred crystallographic orientation along the (111), (200) or (220) crystallographic planes and the crystallite size of the silver nanorods are briefly analyzed.

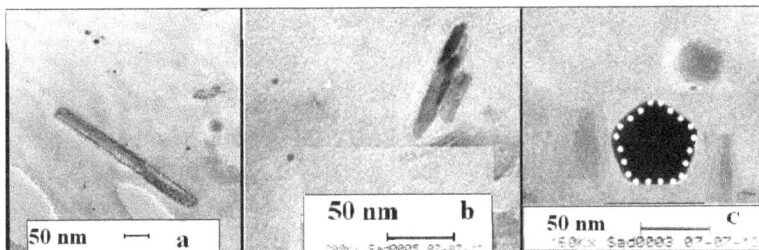

Figure: Transmission electron microscopy (TEM) images of (a,b) individual and (c) cross section of Ag/PVA nanorods.

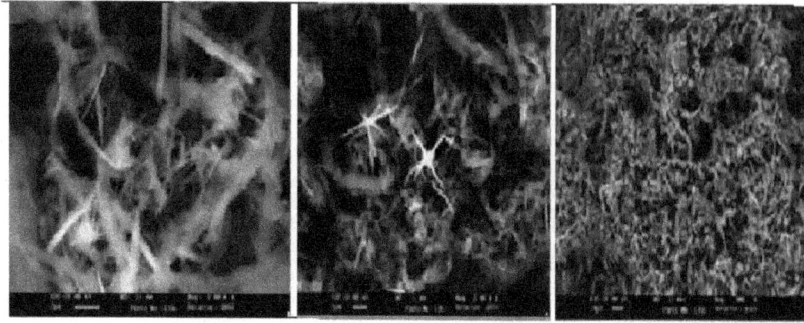

Figure: (a,b,c) SEM image showing high concentrated distribution of Ag/PVA nanorods.

In our research, synthesis of silver nanorods with the controllable dimensions was described by using a reducing agent that involves the reduction of silver nitrate with N-N'-Dimethyl formamide (DMF) in the presence of PVA as a capping reagent. In this process the DMF is served as both reductant and solvent. SEM and TEM observations along a series of relevant directions show that the silver nanorods have an average length of 280 nm and diameters of around 25 nm. TEM observations from cross section of nanorods show that the transformation of silver nanospheres to silver nanorods is achieved by the oriented attachment of several spherical particles followed by their fusion. Resulted Ag-nanorods have a twinned fcc structure, appeared in a pentagonal shape with fivefold twinning. The fivefold axis, i.e. the growth direction, normally goes along the (110) zone axis direction of the fcc cubic structure. The XRD data confirmed that the silver nanorod is crystalline with fcc structure with a preferred crystallographic orientation along the (220) direction and straight, continuous, dense silver nanorod has been obtained with a diameter 25 nm.

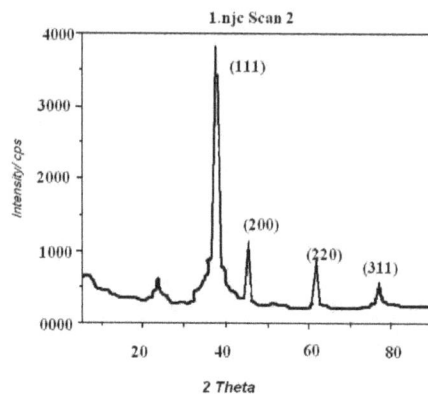

Figure: XRD pattern of Ag/PVA nanocomposite indicative the face-centered cube structure.

On the other hand, gold nanorods were prepared by adopting a photochemical method that employs UV-irradiation to facilitate slow growth of rods Tetraoctylammonium bromide as a co-surfactant used instead of tetradodecylammonium bromide. The growth solution was prepared by dissolving 440 mg of cetyltrimethylammonium bromide (CTAB) and 4.5 mg of tetraoctylammonium bromide (TOAB) in 15 mL of water and transferred to a cylindrical quartz tube (length 15 cm and diameter 2 cm). To this solution, 1.25 mL of 0.024 M $HAuCl_4$ solution was added along with 325 µL of acetone and 225 µL of cyclohexane. The formation of the gold nanorod and its aspect ratio was confirmed from Transmission Electron Microscopic analysis. A drop of a dilute solution of Au nanorods was allowed to dry on a carbon coated copper grid and then probed using a JEOL

JEM-100sx electron microscope. The average length and diameter of rods employed in the present investigation are 50.0 nm and 20.0 nm, respectively and an average aspect ratio of 2.5.

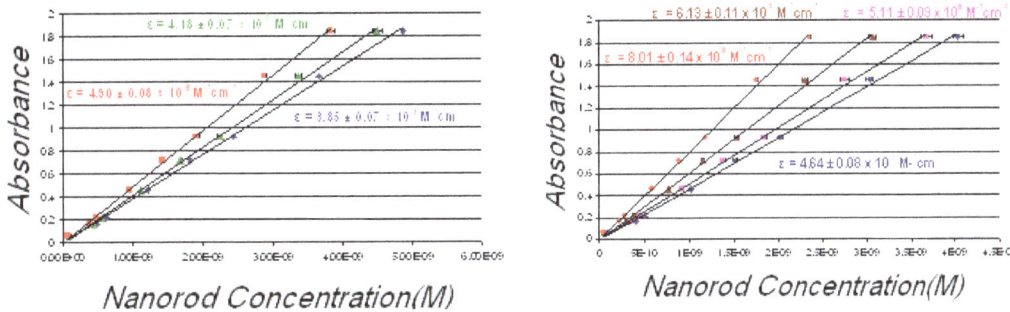

Figure: Effect of error as (a) polydispersity of particles change and (b) number of spherical byproducts increases. In both cases, total gold concentration was held constant. Error bars are very narrow.

Changes in nanorod dimensions have the greatest impact on extinction coefficients. Figure (a) shows the change in the Beer's law plot with increasing polydispersity. As variation in the rod dimensions increases, the extinction coefficient increases. However, these increases are not excessive – from 3.85 ± 0.07 to 4.90 ± 0.08 x 10-8 M is not significant, and errors in concentration resulting from this magnitude of change in ε would not be enough to cause aggregation or substantial miscalculations of antibodies. However, the effect of spherical byproducts is more pronounced, and the presence of spheres can alter the extinction coefficient enough to cause concern.

Figure (b) shows that the extinction coefficient nearly doubles as percentage of spherical by products is increased from 5% to 40%. Spherical byproducts are an ongoing challenge in nanorods synthesis, and it is important to consider how they can change the extinction coefficient.

The ranges of extinction coefficients found in this model are quite different from the values calculated by and , which were on the order of 109 M. However, the rods used in these calculations, while of approximately the same aspect ratio, had dimensions of 50 x 15 nm instead of 30 x 8 nm. The same aspect ratio implies the same peak wavelength, but can result in a different rod volume (and thus different concentration measurements). Small changes in aspect ratio can substantially change absorbance peak location. Literature values must be used cautiously in nanorod studies.

## Application of Nanorod

Nanorods have wide application.

## Nanorods for Dye Solar Cells

The dye-solar cell (DSC or Grätzel) first presented in 1991 offers an interesting paradigm with regards the generation of electricity directly from sunlight via the photovoltaic effect. In essence the DSC is not a definite structure but more a design philosophy to mimicking nature in the conversion of solar energy . The DCS is formed by an electrode, preferably with a large internal surface area onto which is attached a light absorbing dye. The dye upon absorption of a photon by photo-excitation of an electron (moving from the HOMO to the LUMO levels) will, if in favorable conditions, inject the photo-excited electron into the supporting structure. The dye is regenerated by a supporting Reduction & Oxidation (REDOX) electrolyte (or hole conducting semiconductor) which permeates the working electrode. One of the many components of the DSC that can be altered is the working electrode. General requirements are that it be porous (i.e. large internal surface area) and be n-type semiconducting. Several metal-oxides fulfil these requirements (e.g., $TiO_2$, ZnO, $WO_3$).

There are several reports describing the electrodeposition of ZnO with various conditions explored resulting in a variety of geometries as for e.g., nanorods, nanoneedles, nanotubes, nanoporous and compact layers. In this work, a high-density vertically aligned ZnO nanorods arrays, figure, were prepared on fluorine doped $SnO_2$ (FTO) coated glass substrates, prepared at 70 °C from a neutral zinc nitrate solution, varying the deposition time.

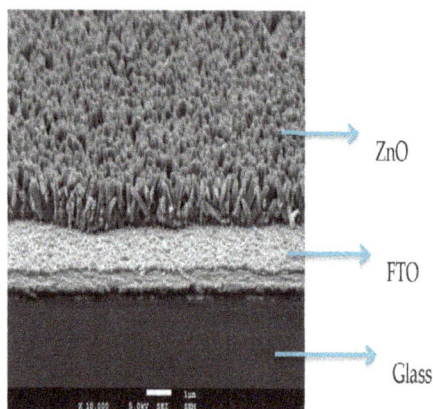

Figure: Cross section view of ZnO nanorod arrays electrodeposited on FTO glass

## The use of Nanorods for Oligonucleotide Detection

Functionalization of nanoparticles with biomolecules (antibodies, nucleic acids, etc.) is of interest for many biomedical applications. One of the most perspectives is application of gold nanorods (GNRs) for detecting target sequences of infecting agents of many dangerous diseases, for example, a HIV-1. The method is based on electrostatic interaction between GNRs and DNA molecules. As single-stranded DNA molecules are zwitterions, and double- stranded are polyanions, their affinity towards polycationic GNR stabilizer (cetiltrimetilammoniumbromide, CTAB) is various. The GNRs plasmon-resonant labels are functionalized with probe single-stranded oligonucleotides by physical adsorption or by chemical attachment. By adding complement targets to the GNRs-probe conjugate solution a formation of aggregates is observed. It has been shown recently that GNRs aggregation, induced by DNA-DNA hybridization on their surface, can be detected by extinction and scattering spectroscopy techniques. We have found that the characteristic parameter of biospecific interaction is change of amplitude and differential light scattering method for detection DNA-DNA interaction. This work is aimed at study of aggregation properties of single-stranded probe DNA-GNRs conjugates as applied to biospecific detection of target oligonucleotides.

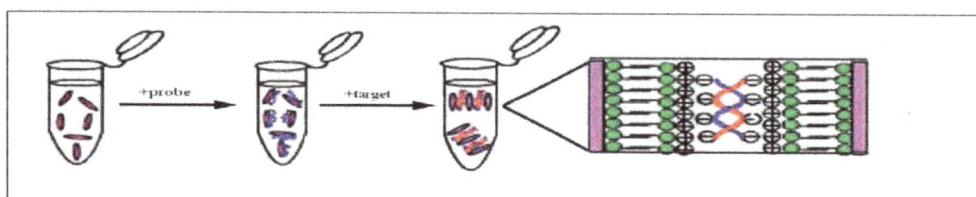

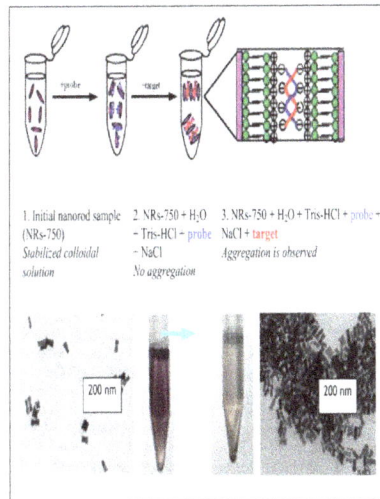

Figure: Experimental diagram for oligonucleotide detection

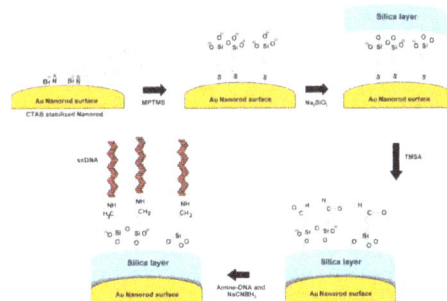

Figure: Schematic diagram of the surface reactions employed to obtain DNA-functionalized gold nanorods.

In above figure, first the nanorods were coated with a thin silica layer using the silane coupling agent, MPTMS, followed by reaction with sodium silicate. Surface-bound aldehyde functional groups were attached to the silica film by using TMSA and then used to conjugate the amine-modified DNA in a reductive amination reaction.

GNRs were produced by seed-mediated growth method in presence of CTAB. Extinction and scattering spectra were measured in the wavelength range from 450 to 900 nanometers which includes the transversal and longitudinal plasmon resonance bands of GNRs. The particle and aggregate average size were measured by the dynamic light scattering method (DLS), and the particle shape and cluster structure were examined by transmission electron microscopy (TEM). 21-mer oligonucleotide complement pair was taken as a biospecific pair, and the target sequence was related to human immunodeficiency virus-1 (HIV-1) genome. In this experiments demonstrated that reproducibility of aggregation test depends on the GNRs synthesis protocol. In particularly, the use of protocol led to unsatisfactory results. According to TEM data, the method realization was successful while using "dog-bone" morphology particles.

The functionalized gold nanorod synthesis can be divided into three main steps:

- Gold Nanorod Fabrication.

- Silica Shell Formation.

- DNA Functionalization

## The use of Nanorods for Applied Electric Field

Prominent among them is in the use in display technologies. By changing the orientation of the nanorods with respect to an applied electric field, the reflectivity of the rods can be altered, resulting in superior displays. Picture quality can be improved radically. Each picture element, known as pixel, is composed of a sharp-tipped device of the scale of a few nanometers. Such TVs, known as field emission TVs, are brighter as the pixels can glow better in every color they take up as they pass through a small potential gap at high currents, emitting electrons at the same time.

Nanorod-based flexible, thin-film computers can revolutionize the retail industry, enabling customers to checkout easily without the hassles of having to pay cash.

The semiconductor nanorod structure is based on a junction between nanorod structure and another window semiconductor layer for solar cell application. The possibility of band gap tuning by varying the diameter of the nanorods along the length, higher absorption coefficient at nanodimensions, the presence of a strong electrical field at the nanorod- window semiconductor nanojunctions and the carrier confinement in lateral direction are expected to result in enhanced absorption and collection efficiency in the proposed device.

## The use of Nanorods for Applied Humidity Sensitive

ZnO nanorod and nanowire films were fabricated on the Si substrates with comb type Pt electrodes by the vapor-phase transport method, and their humidity sensitive characteristics have been investigated. These nanomaterial films show high-humidity sensitivity, good long-term stability and fast response time. It was found that the resistance of the films decreases with increasing relative humidity (RH). At room temperature (RT), resistance changes of more than four and two orders of magnitude were observed when ZnO nanowire and nanorod devices were exposed, respectively, to a moisture pulse of 97% relative humidity. It appears that the ZnO nanomaterial films can be used as efficient humidity sensors. The gas sensor fabricated from ZnO nanorod arrays showed a high sensitivity to $H_2$ from room temperature to a maximum sensitivity at 250 °C and a detection limit of 20 ppm. In addition, the ZnO gas sensor also exhibited excellent responses to NH3 and CO exposure. Our results demonstrate that the hydrothermally grown vertically aligned ZnO nanorod arrays are very promising for the fabrication of cost effective and high performance gas sensors.

To use nanorods in biomedical applications, it is advised that samples are produced in bulk and $\varepsilon$ is calculated for each batch. In addition, Nanorod aggregation, induced by biospecific interaction, was shown by four methods (extinction and scattering spectroscopy, DLS, TEM).

# Quantum Dot

Quantum dots are semiconductor nanoparticles that glow a particular color after being illuminated by light. The color they glow depends on the size of the nanoparticle. When the quantum dots are illuminated by UV light, some of the electrons receive enough energy to break free from the atoms. This capability allows them to move around the nanoparticle, creating a conductance band in which electrons are free to move through a material and conduct electricity. When these electrons drop back into the outer orbit around the atom (the valence band), as illustrated in the following figure, they emit light. The color of that light depends on the energy difference between the conductance band and the valence band.

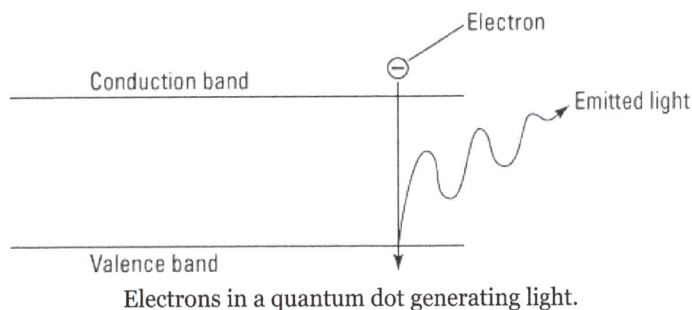

Electrons in a quantum dot generating light.

The smaller the nanoparticle, the higher the energy difference between the valence band and conductance band, which results in a deeper blue color. For a larger nanoparticle, the energy difference between the valence band and the conductance band is lower, which shifts the glow toward red.

Many semiconductor substances can be used as quantum dots, such as cadmium selenide, cadmium sulfide, or indium arsenide. Nanoparticles of these, or any other semiconductor substance, have the properties of a quantum dot. The gap between the valence band and the conductance band, which is present for all semiconductor materials, causes quantum dots to fluoresce.

Quantum dots may be able to increase the efficiency of solar cells. In normal solar cells, a photon of light generates one electron. Experiments with both silicon quantum dots and lead sulfide quantum dots can generate two electrons for a single photon of light. Therefore, using quantum dots in solar cells could significantly increase their efficiency in producing electric power.

Researchers are also working on the use of quantum dots in displays for applications ranging from your cell phone to large screen televisions that would consume less power than current displays. By placing different size quantum dots in each pixel of a display screen, the red, green and blue colors used to generate the full spectrum of colors would be available.

## Production

There are several ways to prepare quantum dots, the principal ones involving colloids.

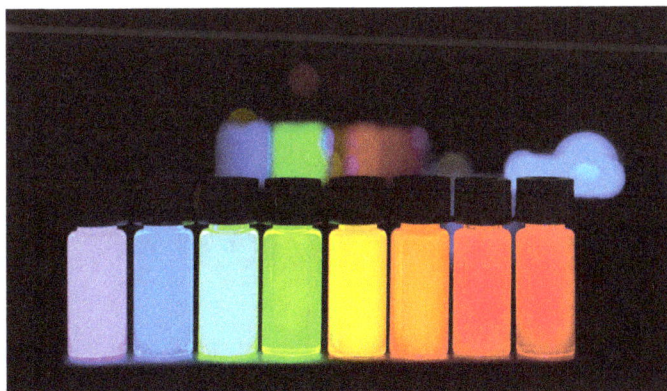

Quantum Dots with gradually stepping emission from violet to deep red.

## Colloidal Synthesis

Colloidal semiconductor nanocrystals are synthesized from solutions, much like traditional chemical processes. The main difference is the product neither precipitates as a bulk solid nor remains dissolved. Heating the solution at high temperature, the precursors decompose forming monomers which then nucleate and generate nanocrystals. Temperature is a critical factor in determining optimal conditions for the nanocrystal growth. It must be high enough to allow for rearrangement and annealing of atoms during the synthesis process while being low enough to promote crystal growth. The concentration of monomers is another critical factor that has to be stringently controlled during nanocrystal growth. The growth process of nanocrystals can occur in two different regimes, "focusing" and "defocusing". At high monomer concentrations, the critical size (the size where nanocrystals neither grow nor shrink) is relatively small, resulting in growth of nearly all particles. In this regime, smaller particles grow faster than large ones (since larger crystals need more atoms to grow than small crystals) resulting in "focusing" of the size distribution to yield nearly monodisperse particles. The size focusing is optimal when the monomer concentration is kept such that the average nanocrystal size present is always slightly larger than the critical size. Over time, the monomer concentration diminishes, the critical size becomes larger than the average size present, and the distribution "defocuses".

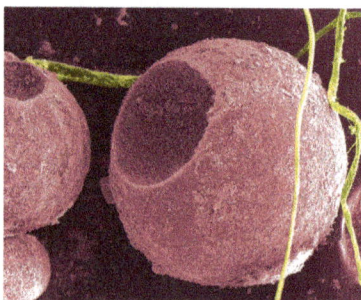

Cadmium sulfide quantum dots on cells.

There are colloidal methods to produce many different semiconductors. Typical dots are made of binary compounds such as lead sulfide, lead selenide, cadmium selenide, cadmium sulfide, cadmium telluride, indium arsenide, and indium phosphide. Dots may also be made from ternary compounds such as cadmium selenide sulfide. These quantum dots can contain as few as 100 to 100,000 atoms within the quantum dot volume, with a diameter of ≈10 to 50 atoms. This corresponds to about 2 to 10 nanometers, and at 10 nm in diameter, nearly 3 million quantum dots could be lined up end to end and fit within the width of a human thumb.

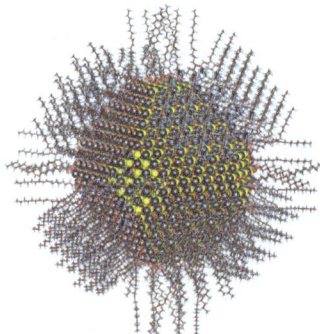

Ideallized image of colloidal nanoparticle of lead sulfide (selenide) with complete passivation by oleic acid, oleyl amine and hydroxyl ligands (size ≈5nm).

Large batches of quantum dots may be synthesized via colloidal synthesis. Due to this scalability and the convenience of benchtop conditions, colloidal synthetic methods are promising for commercial applications. It is acknowledged to be the least toxic of all the different forms of synthesis.

## Plasma Synthesis

Plasma synthesis has evolved to be one of the most popular gas-phase approaches for the production of quantum dots, especially those with covalent bonds. For example, silicon (Si) and germanium (Ge) quantum dots have been synthesized by using nonthermal plasma. The size, shape, surface and composition of quantum dots can all be controlled in nonthermal plasma. Doping that seems quite challenging for quantum dots has also been realized in plasma synthesis. Quantum dots synthesized by plasma are usually in the form of powder, for which surface modification may be carried out. This can lead to excellent dispersion of quantum dots in either organic solvents or water (i. e., colloidal quantum dots).

## Fabrication

- Self-assembled quantum dots are typically between 5 and 50 nm in size. Quantum dots defined by lithographically patterned gate electrodes, or by etching on two-dimensional electron gasses in semiconductor heterostructures can have lateral dimensions between 20 and 100 nm.

- Some quantum dots are small regions of one material buried in another with a larger band gap. These can be so-called core–shell structures, e.g., with CdSe in the core and ZnS in the shell, or from special forms of silica called ormosil. Sub-monolayer shells can also be effective ways of passivating the quantum dots, such as PbS cores with sub-monolayer CdS shells.

- Quantum dots sometimes occur spontaneously in quantum well structures due to monolayer fluctuations in the well's thickness.

- Self-assembled quantum dots nucleate spontaneously under certain conditions during molecular beam epitaxy (MBE) and metallorganic vapor phase epitaxy (MOVPE), when a material is grown on a substrate to which it is not lattice matched. The resulting strain produces coherently strained islands on top of a two-dimensional wetting layer. This growth mode is known as Stranski–Krastanov growth. The islands can be subsequently buried to form the quantum dot. This fabrication method has potential for applications in quantum cryptography (i.e. single photon sources) and quantum computation. The main limitations of this method are the cost of fabrication and the lack of control over positioning of individual dots.

- Individual quantum dots can be created from two-dimensional electron or hole gases present in remotely doped quantum wells or semiconductor heterostructures called lateral quantum dots. The sample surface is coated with a thin layer of resist. A lateral pattern is then defined in the resist by electron beam lithography. This pattern can then be transferred to the electron or hole gas by etching, or by depositing metal electrodes (lift-off process) that allow the application of external voltages between the electron gas and the electrodes. Such quantum dots are mainly of interest for experiments and applications involving electron or hole transport, i.e., an electrical current.

- The energy spectrum of a quantum dot can be engineered by controlling the geometrical size, shape, and the strength of the confinement potential. Also, in contrast to atoms, it is relatively easy to connect quantum dots by tunnel barriers to conducting leads, which allows the application of the techniques of tunneling spectroscopy for their investigation.

The quantum dot absorption features correspond to transitions between discrete, three-dimensional particle in a box states of the electron and the hole, both confined to the same nanometer-size box. These discrete transitions are reminiscent of atomic spectra and have resulted in quantum dots also being called artificial atoms.

- Confinement in quantum dots can also arise from electrostatic potentials (generated by external electrodes, doping, strain, or impurities).

- Complementary metal-oxide-semiconductor (CMOS) technology can be employed to fabricate silicon quantum dots. Ultra small (L=20 nm, W=20 nm) CMOS transistors behave as single electron quantum dots when operated at cryogenic temperature over a range of −269 °C (4 K) to about −258 °C (15 K). The transistor displays Coulomb blockade due to progressive charging of electrons one by one. The number of electrons confined in the channel is driven by the gate voltage, starting from an occupation of zero electrons, and it can be set to 1 or many.

## Viral Assembly

Genetically engineered M13 bacteriophage viruses allow preparation of quantum dot biocomposite structures. It had previously been shown that genetically engineered viruses can recognize specific semiconductor surfaces through the method of selection by combinatorial phage display. Additionally, it is known that liquid crystalline structures of wild-type viruses (Fd, M13, and TMV) are adjustable by controlling the solution concentrations, solution ionic strength, and the external magnetic field applied to the solutions. Consequently, the specific recognition properties of the virus can be used to organize inorganic nanocrystals, forming ordered arrays over the length scale defined by liquid crystal formation. Using this information, Lee were able to create self-assembled, highly oriented, self-supporting films from a phage and ZnS precursor solution. This system allowed them to vary both the length of bacteriophage and the type of inorganic material through genetic modification and selection.

## Electrochemical Assembly

Highly ordered arrays of quantum dots may also be self-assembled by electrochemical techniques. A template is created by causing an ionic reaction at an electrolyte-metal interface which results in the spontaneous assembly of nanostructures, including quantum dots, onto the metal which is then used as a mask for mesa-etching these nanostructures on a chosen substrate.

## Bulk-manufacture

Quantum dot manufacturing relies on a process called "high temperature dual injection" which has been scaled by multiple companies for commercial applications that require large quantities (hundreds of kilograms to tonnes) of quantum dots. This reproducible production method can be applied to a wide range of quantum dot sizes and compositions.

The bonding in certain cadmium-free quantum dots, such as III-V-based quantum dots, is more covalent than that in II-VI materials, therefore it is more difficult to separate nanoparticle nucleation and growth via a high temperature dual injection synthesis. An alternative method of quantum dot synthesis, the "molecular seeding" process, provides a reproducible route to the production of high quality quantum dots in large volumes. The process utilises identical molecules of a molecular cluster compound as the nucleation sites for nanoparticle growth, thus avoiding the need for a high temperature injection step. Particle growth is maintained by the periodic addition of precursors at moderate temperatures until the desired particle size is reached. The molecular seeding process is not limited to the production of cadmium-free quantum dots; for example, the process can be used to synthesise kilogram batches of high quality II-VI quantum dots in just a few hours.

Another approach for the mass production of colloidal quantum dots can be seen in the transfer of the well-known hot-injection methodology for the synthesis to a technical continuous flow system. The batch-to-batch variations arising from the needs during the mentioned methodology can be overcome by utilizing technical components for mixing and growth as well as transport and temperature adjustments. For the production of CdSe based semiconductor nanoparticles this method has been investigated and tuned to production amounts of kg per month. Since the use of technical components allows for easy interchange in regards of maximum through-put and size, it can be further enhanced to tens or even hundreds of kilograms.

In 2011 a consortium of U.S. and Dutch companies reported a "milestone" in high volume quantum dot manufacturing by applying the traditional high temperature dual injection method to a flow system.

On January 23, 2013 Dow entered into an exclusive licensing agreement with UK-based Nanoco for the use of their low-temperature molecular seeding method for bulk manufacture of cadmium-free quantum dots for electronic displays, and on September 24, 2014 Dow commenced work on the production facility in South Korea capable of producing sufficient quantum dots for "millions of cadmium-free televisions and other devices, such as tablets". Mass production is due to commence in mid-2015. On 24 March 2015 Dow announced a partnership deal with LG Electronics to develop the use of cadmium free quantum dots in displays.

## Heavy-metal-free Quantum Dots

In many regions of the world there is now a restriction or ban on the use of heavy metals in many household goods, which means that most cadmium-based quantum dots are unusable for consumer-goods applications.

For commercial viability, a range of restricted, heavy-metal-free quantum dots has been developed showing bright emissions in the visible and near infra-red region of the spectrum and have similar optical properties to those of CdSe quantum dots. Among these systems are InP/ZnS and CuInS/ZnS, for example.

Peptides are being researched as potential quantum dot material.Since peptides occur naturally in all organisms, such dots would likely be nontoxic and easily biodegraded.

## Health and Safety

Some quantum dots pose risks to human health and the environment under certain conditions. Notably, the studies on quantum dot toxicity are focused on cadmium containing particles and has yet to be demonstrated in animal models after physiologically relevant dosing. In vitro studies, based on cell cultures, on quantum dots (QD) toxicity suggests that their toxicity may derive from multiple factors including its physicochemical characteristics (size, shape, composition, surface functional groups, and surface charges) and environment. Assessing their potential toxicity is complex as these factors include properties such as QD size, charge, concentration, chemical composition, capping ligands, and also on their oxidative, mechanical and photolytic stability.

Many studies have focused on the mechanism of QD cytotoxicity using model cell cultures. It has been demonstrated that after exposure to ultraviolet radiation or oxidation by air, CdSe QDs release free cadmium ions causing cell death. Group II-VI QDs also have been reported to induce the formation of reactive oxygen species after exposure to light, which in turn can damage cellular components such as proteins, lipids and DNA. Some studies have also demonstrated that addition of a ZnS shell inhibit the process of reactive oxygen species in CdSe QDs. Another aspect of QD toxicity is the process of their size dependent intracellular pathways that concentrate these particles in cellular organelles that are inaccessible by metal ions, which may result in unique patterns of cytotoxicity compared to their constituent metal ions. The reports of QD localization in the cell nucleus present additional modes of toxicity because they may induce DNA mutation, which in turn will propagate through future generation of cells causing diseases.

Although concentration of QDs in certain organelles have been reported in in vivo studies using animal models, no alterations in animal behavior, weight, hematological markers or organ damage has been found through either histological or biochemical analysis. These finding have led scientists to believe that intracellular dose is the most important deterring factor for QD toxicity. Therefore, factors determining the QD endocytosis that determine the effective intracellular concentration, such as QD size, shape and surface chemistry determine their toxicity. Excretion of QDs through urine in animal models also have demonstrated via injecting radio-labeled ZnS capped CdSe QDs where the ligand shell was labelled with $^{99m}$Tc. Though multiple other studies have concluded retention of QDs in cellular levels, exocytosis of QDs is still poorly studied in the literature.

While significant research efforts have broadened the understanding of toxicity of QDs, there are large discrepancies in the literature and questions still remains to be answered. Diversity of this class material as compared to normal chemical substances makes the assessment of their toxicity very challenging. As their toxicity may also be dynamic depending on the environmental factors such as pH level, light exposure and cell type, traditional methods of assessing toxicity of chemicals such as $LD_{50}$ are not applicable for QDs. Therefore, researchers are focusing on introducing novel approaches and adapting existing methods to include this unique class of materials. Furthermore, novel strategies to engineer safer QDs are still under exploration by the scientific community. A recent novelty in the field is the discovery of carbon quantum dots, a new generation of optically-active nanoparticles potentially capable of replacing semiconductor QDs, but with the advantage of much lower toxicity.

# Optical Properties

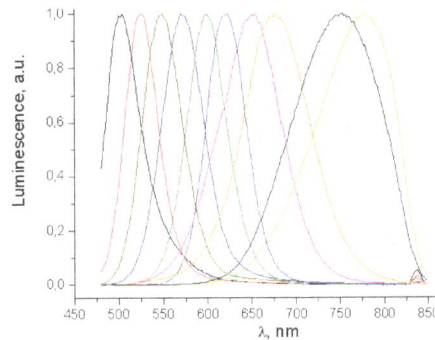

Fluorescence spectra of CdTe quantum dots of various sizes. Different sized quantum dots emit different color light due to quantum confinement.

In semiconductors, light absorption generally leads to an electron being excited from the valence to the conduction band, leaving behind a hole. The electron and the hole can bind to each other to form an exciton. When this exciton recombines (i.e. the electron resumes its ground state), the exciton's energy can be emitted as light. This is called fluorescence. In a simplified model, the energy of the emitted photon can be understood as the sum of the band gap energy between the highest occupied level and the lowest unoccupied energy level, the confinement energies of the hole and the excited electron, and the bound energy of the exciton (the electron-hole pair):

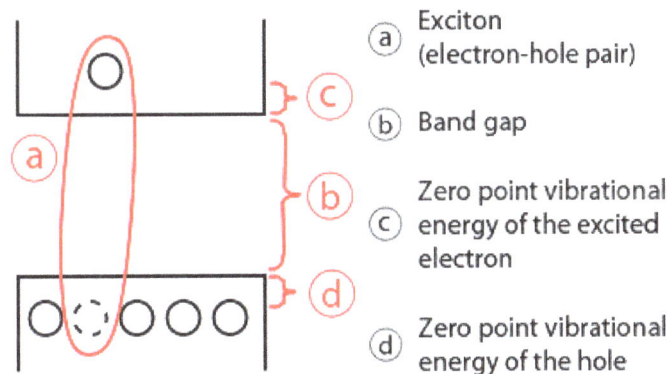

a) Exciton (electron-hole pair)

b) Band gap

c) Zero point vibrational energy of the excited electron

d) Zero point vibrational energy of the hole

As the confinement energy depends on the quantum dot's size, both absorption onset and fluorescence emission can be tuned by changing the size of the quantum dot during its synthesis. The larger the dot, the redder (lower energy) its absorption onset and fluorescence spectrum. Conversely, smaller dots absorb and emit bluer (higher energy) light. Recent articles in Nanotechnology and in other journals have begun to suggest that the shape of the quantum dot may be a factor in the coloration as well, but as yet not enough information is available. Furthermore, it was shown that the lifetime of fluorescence is determined by the size of the quantum dot. Larger dots have more closely spaced energy levels in which the electron-hole pair can be trapped. Therefore, electron-hole pairs in larger dots live longer causing larger dots to show a longer lifetime.

To improve fluorescence quantum yield, quantum dots can be made with "shells" of a larger bandgap semiconductor material around them. The improvement is suggested to be due to the reduced access of electron and hole to non-radiative surface recombination pathways in some cases, but also due to reduced Auger recombination in others.

## Potential Applications

Quantum dots are particularly promising for optical applications due to their high extinction coefficient. They operate like a single electron transistor and show the Coulomb blockade effect. Quantum dots have also been suggested as implementations of qubits for quantum information processing.

Tuning the size of quantum dots is attractive for many potential applications. For instance, larger quantum dots have a greater spectrum-shift towards red compared to smaller dots, and exhibit less pronounced quantum properties. Conversely, the smaller particles allow one to take advantage of more subtle quantum effects.

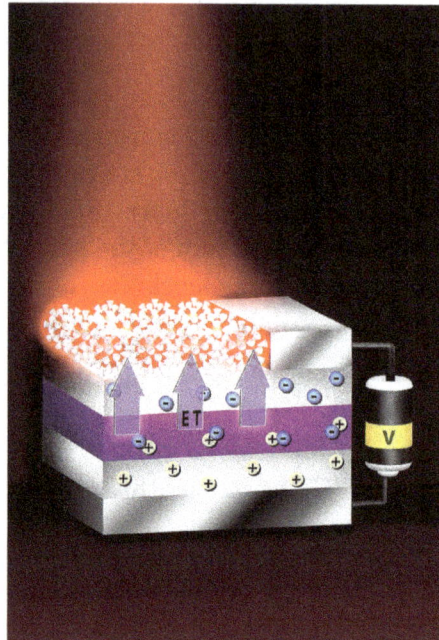

A device that produces visible light, through energy transfer from thin layers of quantum wells to crystals above the layers.

Being zero-dimensional, quantum dots have a sharper density of states than higher-dimensional structures. As a result, they have superior transport and optical properties. They have potential uses in diode lasers, amplifiers, and biological sensors. Quantum dots may be excited within a locally enhanced electromagnetic field produced by gold nanoparticles, which can then be observed from the surface plasmon resonance in the photoluminescent excitation spectrum of $(CdSe)ZnS$ nanocrystals. High-quality quantum dots are well suited for optical encoding and multiplexing applications due to their broad excitation profiles and narrow/symmetric emission spectra. The new generations of quantum dots have far-reaching potential for the study of intracellular processes at the single-molecule level, high-resolution cellular imaging, long-term in vivo observation of cell trafficking, tumor targeting, and diagnostics.

CdSe nanocrystals are efficient triplet photosensitizers. Laser excitation of small CdSe nanoparticles enables the extraction of the excited state energy from the Quantum Dots into bulk solution, thus opening the door to a wide range of potential applications such as photodynamic therapy, photovoltaic devices, molecular electronics, and catalysis.

# Biology

In modern biological analysis, various kinds of organic dyes are used. However, as technology advances, greater flexibility in these dyes is sought. To this end, quantum dots have quickly filled in the role, being found to be superior to traditional organic dyes on several counts, one of the most immediately obvious being brightness (owing to the high extinction coefficient combined with a comparable quantum yield to fluorescent dyes) as well as their stability (allowing much less photobleaching). It has been estimated that quantum dots are 20 times brighter and 100 times more stable than traditional fluorescent reporters. For single-particle tracking, the irregular blinking of quantum dots is a minor drawback. However, there have been groups which have developed quantum dots which are essentially nonblinking and demonstrated their utility in single molecule tracking experiments.

The use of quantum dots for highly sensitive cellular imaging has seen major advances. The improved photostability of quantum dots, for example, allows the acquisition of many consecutive focal-plane images that can be reconstructed into a high-resolution three-dimensional image. Another application that takes advantage of the extraordinary photostability of quantum dot probes is the real-time tracking of molecules and cells over extended periods of time. Antibodies, streptavidin, peptides, DNA, nucleic acid aptamers, or small-molecule ligands can be used to target quantum dots to specific proteins on cells. Researchers were able to observe quantum dots in lymph nodes of mice for more than 4 months.

Semiconductor quantum dots have also been employed for in vitro imaging of pre-labeled cells. The ability to image single-cell migration in real time is expected to be important to several research areas such as embryogenesis, cancer metastasis, stem cell therapeutics, and lymphocyte immunology.

One application of quantum dots in biology is as donor fluorophores in Förster resonance energy transfer, where the large extinction coefficient and spectral purity of these fluorophores make them superior to molecular fluorophores It is also worth noting that the broad absorbance of QDs allows selective excitation of the QD donor and a minimum excitation of a dye acceptor in FRET-based studies. The applicability of the FRET model, which assumes that the Quantum Dot can be approximated as a point dipole, has recently been demonstrated.

The use of quantum dots for tumor targeting under in vivo conditions employ two targeting schemes: active targeting and passive targeting. In the case of active targeting, quantum dots are functionalized with tumor-specific binding sites to selectively bind to tumor cells. Passive targeting uses the enhanced permeation and retention of tumor cells for the delivery of quantum dot probes. Fast-growing tumor cells typically have more permeable membranes than healthy cells, allowing the leakage of small nanoparticles into the cell body. Moreover, tumor cells lack an effective lymphatic drainage system, which leads to subsequent nanoparticle-accumulation.

Quantum dot probes exhibit in vivo toxicity. For example, CdSe nanocrystals are highly toxic to cultured cells under UV illumination, because the particles dissolve, in a process known as photolysis, to release toxic cadmium ions into the culture medium. In the absence of UV irradiation, however, quantum dots with a stable polymer coating have been found to be essentially nontoxic. Hydrogel encapsulation of quantum dots allows for quantum dots to be introduced into a stable aqueous solution, reducing the possibility of cadmium leakage.Then again, only little is known about the excretion process of quantum dots from living organisms.

In another potential application, quantum dots are being investigated as the inorganic fluorophore for intra-operative detection of tumors using fluorescence spectroscopy.

Delivery of undamaged quantum dots to the cell cytoplasm has been a challenge with existing techniques. Vector-based methods have resulted in aggregation and endosomal sequestration of quantum dots while electroporation can damage the semi-conducting particles and aggregate delivered dots in the cytosol. Via cell squeezing, quantum dots can be efficiently delivered without inducing aggregation, trapping material in endosomes, or significant loss of cell viability. Moreover, it has shown that individual quantum dots delivered by this approach are detectable in the cell cytosol, thus illustrating the potential of this technique for single molecule tracking studies.

## Photovoltaic Devices

The tunable absorption spectrum and high extinction coefficients of quantum dots make them attractive for light harvesting technologies such as photovoltaics. Quantum dots may be able to increase the efficiency and reduce the cost of today's typical silicon photovoltaic cells. According to an experimental proof from 2004, quantum dots of lead selenide can produce more than one exciton from one high energy photon via the process of carrier multiplication or multiple exciton generation (MEG). This compares favorably to today's photovoltaic cells which can only manage one exciton per high-energy photon, with high kinetic energy carriers losing their energy as heat. Quantum dot photovoltaics would theoretically be cheaper to manufacture, as they can be made "using simple chemical reactions."

### Quantum Dot only Solar Cells

Aromatic self-assembled monolayers (SAMs) (e.g. 4-nitrobenzoic acid) can be used to improve the band alignment at electrodes for better efficiencies. This technique has provided a record power conversion efficiency (PCE) of 10.7%. The SAM is positioned between ZnO-PbS colloidal quantum dot (CQD) film junction to modify band alignment via the dipole moment of the constituent SAM molecule, and the band tuning may be modified via the density, dipole and the orientation of the SAM molecule.

### Quantum Dot in Hybrid Solar Cells

Colloidal quantum dots are also used in inorganic/organic hybrid solar cells. These solar cells are attractive because of the potential for low-cost fabrication and relatively high efficiency. Incorporation of metal oxides, such as ZnO, TiO2, and Nb2O5 nanomaterials into organic photovoltaics have been commercialized using full roll-to-roll processing. A 13.2% power conversion efficiency is claimed in Si nanowire/PEDOT:PSS hybrid solar cells.

### Quantum Dot with Nanowire in Solar Cells

Another potential use involves capped single-crystal ZnO nanowires with CdSe quantum dots, immersed in mercaptopropionic acid as hole transport medium in order to obtain a QD-sensitized solar cell. The morphology of the nanowires allowed the electrons to have a direct pathway to the photoanode. This form of solar cell exhibits 50–60% internal quantum efficiencies.

Nanowires with quantum dot coatings on silicon nanowires (SiNW) and carbon quantum dots. The use of SiNWs instead of planar silicon enhances the antiflection properties of Si. The SiNW exhibits a light-trapping effect due to light trapping in the SiNW. This use of SiNWs in conjunction with carbon quantum dots resulted in a solar cell that reached 9.10% PCE.

Graphene quantum dots have also been blended with organic electronic materials to improve efficiency and lower cost in photovoltaic devices and organic light emitting diodes (OLEDs) in compared to graphene sheets. These graphene quantum dots were functionalized with organic ligands that experience photoluminescence from UV-Vis absorption.

## Light Emitting Diodes

Several methods are proposed for using quantum dots to improve existing light-emitting diode (LED) design, including "Quantum Dot Light Emitting Diode" (QD-LED or QLED) displays and "Quantum Dot White Light Emitting Diode" (QD-WLED) displays. Because Quantum dots naturally produce monochromatic light, they can be more efficient than light sources which must be color filtered. QD-LEDs can be fabricated on a silicon substrate, which allows them to be integrated onto standard silicon-based integrated circuits or microelectromechanical systems.

## Quantum Dot Displays

Quantum dots are valued for displays, because they emit light in very specific gaussian distributions. This can result in a display with visibly more accurate colors. A conventional color liquid crystal display (LCD) is usually backlit by fluorescent lamps (CCFLs) or conventional white LEDs that are color filtered to produce red, green, and blue pixels. An improvement is using conventional blue-emitting LEDs as the light sources and converting part of the emitted light into *pure* green and red light by the appropriate quantum dots placed in front of the blue LED or using a quantum dot infused diffuser sheet in the backlight optical stack. This type of white light as the backlight of an LCD panel allows for the best color gamut at lower cost than a RGB LED combination using three LEDs.

The ability of QDs to precisely convert and tune a spectrum makes them attractive for LCD displays. Previous LCD displays can waste energy converting red-green poor, blue-yellow rich white light into a more balanced lighting. By using QDs, only the necessary colors for ideal images are contained in the screen. The result is a screen that is brighter, clearer, and more energy-efficient. The first commercial application of quantum dots was the Sony XBR X900A series of flat panel televisions released in 2013.

In June 2006, QD Vision announced technical success in making a proof-of-concept quantum dot display and show a bright emission in the visible and near infra-red region of the spectrum. A QD-LED integrated at a scanning microscopy tip was used to demonstrate fluorescence near-field scanning optical microscopy (NSOM) imaging.

## Photodetector Devices

Quantum dot photodetectors (QDPs) can be fabricated either via solution-processing, or from conventional single-crystalline semiconductors. Conventional single-crystalline semiconductor QDPs are precluded from integration with flexible organic electronics due to the incompatibility of their growth conditions with the process windows required by organic semiconductors. On the other hand, solution-processed QDPs can be readily integrated with an almost infinite variety of substrates, and also postprocessed atop other integrated circuits. Such colloidal QDPs have potential applications in surveillance, machine vision, industrial inspection, spectroscopy, and fluorescent biomedical imaging.

## Photocatalysts

Quantum dots also function as photocatalysts for the light driven chemical conversion of water into hydrogen as a pathway to solar fuel. In photocatalysis, electron hole pairs formed in the dot under band gap excitation drive redox reactions in the surrounding liquid. Generally, the photocatalytic activity of the dots is related to the particle size and its degree of quantum confinement. This is because the band gap determines the chemical energy that is stored in the dot in the excited state. An obstacle for the use of quantum dots in photocatalysis is the presence of surfactants on the surface of the dots. These surfactants (or ligands) interfere with the chemical reactivity of the dots by slowing down mass transfer and electron transfer processes. Also, quantum dots made of metal chalcogenides are chemically unstable under oxidizing conditions and undergo photo corrosion reactions.

## Theory

Quantum dots are theoretically described as a point like, or a zero dimensional (0D) entity. Most of their properties depend on the dimensions, shape and materials of which QDs are made. Generally QDs present different thermodynamic properties from the bulk materials of which they are made. One of these effects is the Melting-point depression. Optical properties of spherical metallic QDs are well described by the Mie scattering theory.

## Quantum Confinement in Semiconductors

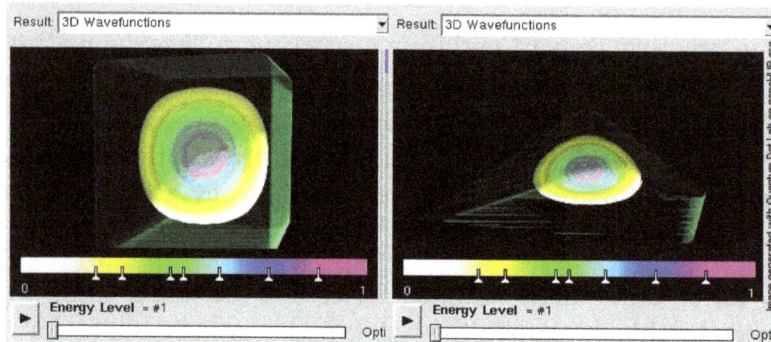

3D confined electron wave functions in a quantum dot. Here, rectangular and triangular-shaped quantum dots are shown. Energy states in rectangular dots are more *s-type* and *p-type*. However, in a triangular dot the wave functions are mixed due to confinement symmetry.

In a semiconductor crystallite whose size is smaller than twice the size of its exciton Bohr radius, the excitons are squeezed, leading to quantum confinement. The energy levels can then be predicted using the particle in a box model in which the energies of states depend on the length of the box. Comparing the quantum dots size to the Bohr radius of the electron and hole wave functions, 3 regimes can be defined. A 'strong confinement regime' is defined as the quantum dots radius being smaller than both electron and hole Bohr radius, 'weak confinement' is given when the quantum dot is larger than both. For semiconductors in which electron and hole radii are markedly different, an 'intermediate confinement regime' exists, where the quantum dot's radius is larger than the Bohr radius of one charge carrier (typically the hole), but not the other charge carrier.

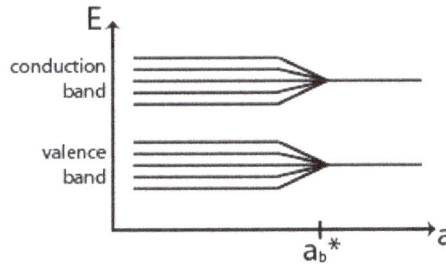

Splitting of energy levels for small quantum dots due to the quantum confinement effect. The horizontal axis is the radius, or the size, of the quantum dots and $a_b$* is the Exciton Bohr radius.

## Band Gap Energy

The band gap can become smaller in the strong confinement regime as the energy levels split up. The Exciton Bohr radius can be expressed as:

$$a_b^* = \varepsilon_r \left( \frac{m}{\mu} \right) a_b$$

where $a_b$ is the Bohr radius=0.053 nm, m is the mass, $\mu$ is the reduced mass, and $\varepsilon_r$ is the size-dependent dielectric constant (Relative permittivity). This results in the increase in the total emission energy (the sum of the energy levels in the smaller band gaps in the strong confinement regime is larger than the energy levels in the band gaps of the original levels in the weak confinement regime) and the emission at various wavelengths. If the size distribution of QDs is not enough peaked, the convolution of multiple emission wavelengths is observed as a continuous spectra.

## Confinement energy

The exciton entity can be modeled using the particle in the box. The electron and the hole can be seen as hydrogen in the Bohr model with the hydrogen nucleus replaced by the hole of positive charge and negative electron mass. Then the energy levels of the exciton can be represented as the solution to the particle in a box at the ground level (n = 1) with the mass replaced by the reduced mass. Thus by varying the size of the quantum dot, the confinement energy of the exciton can be controlled.

## Bound Exciton Energy

There is Coulomb attraction between the negatively charged electron and the positively charged hole. The negative energy involved in the attraction is proportional to Rydberg's energy and inversely proportional to square of the size-dependent dielectric constant of the semiconductor.

When the size of the semiconductor crystal is smaller than the Exciton Bohr radius, the Coulomb interaction must be modified to fit the situation.

Therefore, the sum of these energies can be represented as:

$$E_{confinement} = \frac{\hbar^2 \pi^2}{2a^2} \left( \frac{1}{m_e} + \frac{1}{m_h} \right) = \frac{\hbar^2 \pi^2}{2\mu a^2}$$

$$E_{exciton} = -\frac{1}{\epsilon_r^2} \frac{\mu}{m_e} R_y = -R_y^*$$

$$E = E_{bandgap} + E_{confinement} + E_{exciton}$$

$$= E_{bandgap} + \frac{\hbar^2 \pi^2}{2\mu a^2} - R_y^*$$

where $\mu$ is the reduced mass, $a$ is the radius of the quantum dot, $m_e$ is the free electron mass, $m_h$ is the hole mass, and $\varepsilon_r$ is the size-dependent dielectric constant.

Although the above equations were derived using simplifying assumptions, they imply that the electronic transitions of the quantum dots will depend on their size. These quantum confinement effects are apparent only below the critical size. Larger particles do not exhibit this effect. This effect of quantum confinement on the quantum dots has been repeatedly verified experimentally and is a key feature of many emerging electronic structures.

The Coulomb interaction between confined carriers can also be studied by numerical means when results unconstrained by asymptotic approximations are pursued.

Besides confinement in all three dimensions (i.e., a quantum dot), other quantum confined semi-conductors include:

- Quantum wires, which confine electrons or holes in two spatial dimensions and allow free propagation in the third.

- Quantum wells, which confine electrons or holes in one dimension and allow free propagation in two dimensions.

## Nanocrystalline Material

Nanocrystalline materials are single- or multi-phase polycrystalline solids with a grain size of a few nanometers (1 nm = 10−9 m = 10 Å), typically less than 100 nm. Since the grain sizes are so small, a significant volume of the microstructure in nanocrystalline materials is composed of interfaces, mainly grain boundaries, i.e., a large volume fraction of the atoms resides in grain boundaries. Consequently, nanocrystalline materials exhibit properties that are significantly different from and often improved over, their conventional coarse-grained polycrystalline counterparts.

Nanocrystalline phases were detected in samples of lunar soils. Many conventional catalytic materials are based on very fine microstructures. Nanostructures formed chemically under ambient conditions can also be found in natural biological systems from seashells to bone and teeth in the human body. These materials are notable in that they are simultaneously hard, strong, and tough. Therefore, a number of investigations have been conducted to mimic nature (biomimetics) and also artificially synthesize nanostructured materials and study their properties and behavior. These investigations have clearly shown that one could engineer (tailor) the properties of nanocrystalline materials through control of microstructural features, more specifically the grain size.

## Classification

Nanocrystalline materials can be classified into different categories depending on the number of dimensions in which the material has nanometer modulations. Thus, they can be classified into (a) layered or lamellar structures, (b) filamentary structures, and (c) equiaxed nanostructured materials. A layered or lamellar structure is a onedimensional (1D) nanostructure in which the magnitudes of length and width are much greater than the thickness that is only a few nanometers in size.

Table: Classification of nanocrystalline materials

| Dimensionality | Designation | Typical method(s) of synthesis |
|---|---|---|
| One-dimensional (1D) | Layered (lamellar) | Vapor deposition |
| | | Electrodeposition |
| Two-dimensional (2D) | Filamentary | Chemical vapor deposition |
| Three-dimensional (3D) | Crystallites | Gas condensation |
| | (equiaxed) | Mechanical alloying/milling |

One can also visualize a two-dimensional (2D) rod-shaped nanostructure that can be termed filamentary and in this the length is substantially larger than width or diameter, which are of nanometer dimensions. The most common of the nanostructures, however, is basically equiaxed (all the three dimensions are of nanometer size) and are termed nanostructured crystallites (three-dimensional [3D] nanostructures).

The nanostructured materials may contain crystalline, quasicrystalline, or amorphous phases and can be metals, ceramics, polymers, or composites. If the grains are made up of crystals, the material is called nanocrystalline. On the other hand, if they are made up of quasicrystalline or amorphous (glassy) phases, they are termed nanoquasicrystals and nanoglasses, respectively.

## Synthesis

Nanocrystalline materials can be synthesized either by consolidating small clusters or breaking down the bulk material into smaller and smaller dimensions. Gleiter used the inert gas condensation technique to produce nanocrystalline powder particles and consolidated them in situ into small disks under ultra-high vacuum (UHV) conditions. Since then a number of techniques have been developed to prepare nanostructured materials starting from the vapor, liquid, or solid states. Table lists some of the more common methods used to produce nanocrystalline materials and also the dimensionality of the product obtained.

Table: Methods to synthesize nanocrystalline materials.

| Starting Phase | Technique | Dimensionality of Produc |
|---|---|---|
| Vapor | Inert gas condensation | 3D |
| | Physical vapor deposition – Evaporation and sputtering | 1D |
| | Plasma processing | 3D |
| | Chemical vapor condensation | 3D, 2D |
| | Chemical reactions | 3D |
| Liquid | Rapid solidification | 1D, 3D |
| | Electrodeposition | 3D |
| | Chemical reactions | 3D |
| Solid | Mechanical alloying/milling | 3D |
| | Devitrification of amorphous phases | 3D |
| | Spark erosion | 3D |
| | Sliding wear | 3D |

Nanostructured materials have been synthesized in recent years by methods including inert gas condensation, mechanical alloying, spray conversion processing, severe plastic deformation, electro deposition, rapid solidification from the melt, physical vapor deposition, chemical vapor processing, co-precipitation, sol-gel processing, sliding wear, spark erosion, plasma processing, auto-ignition, laser ablation, hydrothermal pyrolysis, thermophoretic forced flux system, quenching the melt under high pressure, biological templating, sonochemical synthesis, and devitrification of amorphous phases. Actually, in practice any method capable of producing very fine grain-sized materials can be used to synthesize nanocrystalline materials. The grain size, morphology, and texture can be varied by suitably modifying/controlling the process variables in these methods. Each of these methods has advantages and disadvantages and one should choose the appropriate method depending upon the requirements. If a phase transformation is involved, e.g., liquid to solid or vapor to solid, then steps need to be taken to increase the nucleation rate and decrease the growth rate during formation of the product phase. In fact, it is this strategy that is used during devitrification of metallic glasses to produce nanocrystalline materials.

The choice of the method depends upon the ability to control the most important feature of the nanocrystalline materials, viz., the microstructural features (grain size, layer spacing, etc.). Other aspects of importance are the chemical composition and surface chemistry or cleanliness of the interfaces. Extremely clean interfaces can be produced and retained during processing and subsequent consolidation by conducting the experiments under UHV conditions; but, this adds up to the cost of processing. On the other hand, there are also methods that can be very inexpensive; but the purity of the product may not be high. Inert gas condensation, mechanical alloying/milling, spray conversion processing, electrodeposition, and devitrification of amorphous phases are some of the more popular techniques used to produce nanocrystalline materials.

Vapor condensation has been known to produce very fine-grained or amorphous alloys depending on the substrate temperature and other operating conditions. Thus, this technique was originally used to synthesize small quantities of nanostructured pure metals. A number of variants have also been subsequently developed.

The inert gas condensation technique, popularized by Gleiter, consists of evaporating a metal (by resistive heating, radio-frequency heating, sputtering, electron beam heating, laser/plasma heating, or ion sputtering) inside a chamber that was evacuated to a very high vacuum of about $10^{-7}$ torr and then backfilled with a low pressure inert gas, typically a few hundred pascals of helium. The evaporated atoms collide with the gas atoms inside the chamber, lose their kinetic energy, and condense in the form of small, discrete crystals of loose powder. Convection currents, generated due to the heating of the inert gas by the evaporation source and cooled by the liquid nitrogen-filled collection device (cold finger), carry the condensed fine powders to the collector device, from where they can be stripped off by moving an annular teflon ring down the length of the tube into a compaction device. Compaction is carried out in a two-stage piston-and-anvil device initially at low pressures in the upper chamber to produce a loosely compacted pellet, which is then transferred in the vacuum system to a high-pressure unit where final compaction takes place. The scraping and compaction processes also are carried out under UHV conditions to maintain cleanliness of the particle surfaces (and subsequent interfaces) and also to minimize the amount of any trapped gases.

The inert gas condensation method produces equiaxed (3D) crystallites. The crystal size of the powder is typically a few nanometers and the size distribution is narrow. The crystal size is dependent upon the inert gas pressure, the evaporation rate, and the gas composition. Extremely fine particles can be produced by decreasing either the gas pressure in the chamber or the evaporation rate and by using light (such as He) rather than heavy inert gases (such as Xe).

Nanocrystalline alloys can be synthesized by evaporating the different metals from more than one evaporation source. Rotation of the cold finger helps in achieving a better mixing of the vapor. Oxides, nitrides, carbides, etc. of the metals can be synthesized by filling the chamber with oxygen or nitrogen gases or by maintaining a carbonaceous atmosphere. Additionally, at small enough particle sizes, metastable phases are also produced. Thus, this method allows the synthesis of a variety of nanocrystalline materials. High densities of as-compacted samples have been measured with values of about 75–90% of bulk density for metal samples.

## Mechanical Alloying

Mechanical alloying produces nanostructured materials by the structural disintegration of coarser-grained structure as a result of severe plastic deformation. Mechanical alloying consists of repeated welding, fracturing, and rewelding of powder particles in a dry high-energy ball mill until the composition of the resultant powder corresponds to the percentages of the respective constituents in the initial charge. In this process, mixtures of elemental or prealloyed powders are subjected to grinding under a protective atmosphere in equipment capable of high-energy compressive impact forces such as attrition mills, vibrating ball mills, and shaker mills. A majority of the work on nanocrystalline materials has been carried out in highly energetic small shaker mills. The process is referred to as mechanical alloying when one starts with a blended mixture of elemental powders and as mechanical milling when one starts with single component powders such as elements or intermetallic compounds. While material transfer is involved in mechanical alloying, no material transfer is involved in mechanical milling. These processes have produced nanocrystalline structures in pure metals, intermetallic compounds, and immiscible alloy systems. It has been shown that nanometer-sized grains can be obtained in almost any material after sufficient milling time. The grain sizes were found to decrease with milling time down to a minimum value that appeared to scale inversely with the melting temperature.

## Spray Conversion Processing

This process starts with aqueous solution precursors such as ammonium metatungstate [$(NH_4)_6(H_2W_{12}O_{40})\cdot 4H_2O$] and $CoCl_2$, [$Co(CH_3COO)_2$], or cobalt nitrate [$Co(NO_3)_2$]. The solution mixture is aerosolized and rapidly spray dried to give extremely fine mixtures of tungsten and cobalt complex compounds. This precursor powder is then reduced with hydrogen and reacted with carbon monoxide in a fluidized-bed reactor to yield nanophase cobalt/tungsten carbide powder. The tungsten particles are 20–40 nm in size. A typical powder particle consists of a hollow, porous 75 μm sphere containing hundreds of millions of WC grains in a cobalt matrix. To prevent grain growth of tungsten, additions of inhibitors such as VC and $Cr_3C_2$ are made as binders during the sintering steps. Recently, vanadium is being introduced into the starting solution itself to achieve a more uniform distribution in the powder mixture. The process parameters are being further optimized and since the process is fully integrated, on-line control is also being planned.

## Electrodeposition

This is a simple and well-established process and can be easily adapted to produce nanocrystalline materials. Electrodeposition of multilayered (1D) metals can be achieved using either two separate electrolytes or much more conveniently from one electrolyte by appropriate control of agitation and the electrical conditions (particularly voltage). Also, 3D nanostructure crystallites can be prepared using this method by utilizing the interference of one ion with the deposition of the other. It has been shown that electrodeposition yields grain sizes in the nanometer range when the electrodeposition variables (e.g., bath composition, pH, temperature, current density, etc.) are chosen such that nucleation of new grains is favored rather than growth of existing grains. This was achieved by using high deposition rates, formation of appropriate complexes in the bath, addition of suitable surface-active elements to reduce surface diffusion of ad-atoms, etc. This technique can yield porosity-free finished products that do not require subsequent consolidation processing. Further, the process requires low initial capital investment and provides high production rates with few shape and size limitations.

## Devitrification of Amorphous Phases

Many non-equilibrium processing techniques such as rapid solidification from the liquid state, mechanical alloying/milling, electro deposition, and vapor deposition can produce amorphous (glassy) alloys. Controlled crystallization of these amorphous alloys (by increasing the rate of nucleation and decreasing the rate of growth) leads to the synthesis of nanostructured materials. In fact, the most common method to produce nanocrystalline magnetic materials has been to obtain an amorphous phase by rapidly solidifying the melt of appropriate composition and then crystallizing the glassy phase at a relatively low temperature. These materials – referred to as FINEMET – were first investigated by Yoshizawa and this technique has now become an established practice to study the structure and properties of nanocrystalline magnetic materials. This simple devitrification method has been commonly employed to study the magnetic properties of nanocrystalline materials because it can produce (a) porosity-free samples, (b) samples with different grain sizes by controlling the crystallization parameters, and (c) large quantities of material. Furthermore, since no artificial consolidation process is involved (like in inert gas condensation, mechanical alloying, or plasma processing, where fine powders are produced), the interfaces are clean and the product is dense. Additionally, samples with different grain sizes can be synthesized by controlling the crystallization process, affording a way of comparing the properties of amorphous, nanocrystalline, and coarse-grained materials of the same composition.

## Consolidation of Powder to Bulk Shapes

Widespread application of nanocrystalline materials requires production of the powder in tonnage quantities and also efficient methods of consolidating the powders into bulk shapes. The product of majority of the methods to synthesize nanocrystalline materials described above (with the exception of electrodeposition and devitrification of amorphous phases) is powder and therefore this needs to be consolidated. All the consolidation methods generally used in powder metallurgy processes can also be used for nanocrystalline materials. However, because of the small size of the powder particles (typically a few microns, even though the grain size is only a few nanometers), some special precautions need to be taken to minimize their chemical activity and also the high level of interparticle friction.

Successful consolidation of nanocrystalline powders is a non-trivial problem since fully dense materials should be produced while simultaneously retaining the nanometersized grains without coarsening. Conventional consolidation of powders to full density through processes such as hot extrusion and hot isostatic pressing requires use of high pressures and elevated temperatures for extended periods of time to achieve full densification. Unfortunately, however, this results in significant coarsening of the nanometersized grains and consequently the benefits of nanostructure processing are lost. On the other hand, retention of nanostructures requires use of low consolidation temperatures and it is difficult to achieve full inter-particle bonding at these low temperatures. Therefore, novel and innovative methods of consolidating nanocrystalline powders are required.

Successful consolidation of nanocrystalline powders has been achieved by electrodischarge compaction, plasma-activated sintering, shock (explosive) consolidation, hot-isostatic pressing (HIP), Ceracon processing (the Ceracon process (CERAmic CONsolidation) involves taking a heated preform and consolidating the material by pressure against a granular ceramic medium using a conventional forging press), hydrostatic extrusion, strained powder rolling, and sinter forging. By utilizing the combination of high temperature and pressure, HIP can achieve a particular density at lower pressure when compared to cold isostatic pressing or at lower temperature when compared to sintering.

## References

- Delerue, C. & Lannoo, M. (2004). Nanostructures: Theory and Modelling. Springer. p. 47. ISBN 978-3-540-20694-1

- Structural-characterization, nanoparticles-and-nanomaterials: nppt.uni-due.de, Retrieved 29 April 2018

- Dahan, M. (2003). "Diffusion Dynamics of Glycine Receptors Revealed by Single-Quantum Dot Tracking". Science. 302 (5644): 442–5. Bibcode:2003Sci...302..442D. doi:10.1126/science.1088525. PMID 14564008

- Crystal-structures-cross-cutting-topics-1472: nanopartikel.info, Retrieved 12 March 2018

- MFTTech (24 March 2015). "LG Electronics Partners with Dow to Commercialize LGs New Ultra HD TV with Quantum Dot Technology". Retrieved 9 May 2015

- Nanotechnology-photovoltaic-energy-19856: igi-global.com, Retrieved 28 May 2018

- Leatherdale, C. A.; Woo, W. -K.; Mikulec, F. V.; Bawendi, M. G. (2002). "On the Absorption Cross Section of CdSe Nanocrystal Quantum Dots". The Journal of Physical Chemistry B. 106 (31): 7619–7622. doi:10.1021/jp025698c

- Synthesis-and-application-of-nanorods: intechopen.com, Retrieved 15 June 2018

- Juan Carlos Stockert, Alfonso Blázquez-Castro (2017). "Chapter 18 Luminescent Solid-State Markers". Fluorescence Microscopy in Life Sciences. Bentham Science Publishers. pp. 606–641. ISBN 978-1-68108-519-7. Retrieved 24 December 2017

- Quantum-dots: understandingnano.com, Retrieved 10 May 2018

- Ballou, B.; Lagerholm, B. C.; Ernst, L. A.; Bruchez, M. P.; Waggoner, A. S. (2004). "Noninvasive Imaging of Quantum Dots in Mice". Bioconjugate Chemistry. 15 (1): 79–86. doi:10.1021/bc034153y. PMID 14733586

# 4

# Nanosynthesis

The objective of nanosynthesis is to produce a nanomaterial that exhibits desirable properties relative to the length scale of the materials. Accordingly, the synthesis should display control of size within the nano scale. Such methods of nanosynthesis are classified as bottom up and top down. The different aspects of nanoparticle synthesis, mechanical attrition, pyrolysis, etc. have been discussed in this chapter.

## Nanomaterials Synthesis

### Introduction to Synthesis of Nanomaterials

Materials scientists are conducting research to develop novel materials with better properties, more functionality and lower cost than the existing one. Several physical, chemical methods have been developed to enhance the performance of nanomaterials displaying improved properties with the aim to have a better control over the particle size, distribution.

### Methods to Synthesis of Nanomaterials

In general, top-down and bottom-up are the two main approaches for nanomaterials synthesis:

    a. Top-down: size reduction from bulk materials.

    b. Bottom-up: material synthesis from atomic level.

Top-down routes are included in the typical solid –state processing of the materials. This route is based with the bulk material and makes it smaller, thus breaking up larger particles by the use of physical processes like crushing, milling or grinding. Usually this route is not suitable for preparing uniformly shaped materials, and it is very difficult to realize very small particles even with high energy consumption. The biggest problem with top-down approach is the imperfection of the surface structure. Such imperfection would have a significant impact on physical properties and surface chemistry of nanostructures and nanomaterials. It is well known that the conventional top-down technique can cause significant crystallographic damage to the processed patterns.

Bottom up approach refers to the build-up of a material from the bottom: atom-by-atom, molecule-by-molecule or cluster-by-cluster. This route is more often used for preparing most of the nano-scale materials with the ability to generate a uniform size, shape and distribution. It effectively covers chemical synthesis and precisely controlled the reaction to inhibit further particle growth. Although the bottom-up approach is nothing new, it plays an important role in the fabrication and processing of nanostructures and nanomaterials.

Synthesis of nanoparticles to have a better control over particles size distribution, morphology, purity, quantity and quality, by employing environment friendly economical processes has always been a challenge for the researchers. The choice of synthesis technique can be a key factor in determining the effectiveness of the photovoltaic as studies. There are many methods of synthesizing titanium dioxide, such as hydrothermal, combustion synthesis, gas-phase methods, microwave synthesis and sol-gel processing. This research focuses on solgel processing and characterization techniques which was discussed in great detail.

## Hydrothermal Synthesis

Hydrothermal synthesis is typically carried out in a pressurised vessel called an autoclave with the reaction in aqueous solution. The temperature in the autoclave can be raised above the boiling point of water, reaching the pressure of vapour saturation. Hydrothermal synthesis is widely used for the preparation of $TiO_2$ nanoparticles which can easily be obtained through hydrothermal treatment of peptised precipitates of a titanium precursor with water. The hydrothermal method can be useful to control grain size, particle morphology, crystalline phase and surface chemistry through regulation of the solution composition, reaction temperature, pressure, solvent properties, additives and aging time.

## Solvothermal Method

The Solvothermal method is identical to the hydrothermal method except that a variety of solvents other than water can be used for this process. This method has been found to be a versatile route for the synthesis of a wide variety of nanoparticles with narrow size distributions, particularly when organic solvents with high boiling points are chosen. The solvothermal method normally has better control of the size and shape distributions and the crystallinity than the hydrothermal method, and has been employed to synthesize $TiO_2$ nanoparticles and nanorods with/without the aid of surfactants.

## Chemical Vapor Deposition (CVD)

This process is often used in the semiconductor industry to produce high-purity, high-performance thin films. In a typical CVD process, the substrate is exposed to volatile precursors, which react and/ or decompose on the substrate surface to produce the desired film. Frequently, volatile by products that are produced are removed by gas flow through the reaction chamber. The quality of the deposited materials strongly depends on the reaction temperature, the reaction rate, and the concentration of the precursors. Cao prepared $Sn_4{}^+$-doped $TiO_2$ nanoparticles films by the CVD method and found that more surface defects were present on the surface due to doping with Sn. Gracia synthesized M (Cr, V, Fe, Co)-doped TiO2 by CVD and found that TiO2 crystallized into the anatase or rutile structures depending on the type and amount of cation present in the synthesis process. Moreover, upon annealing, partial segregation of the cations in the form of $M_2O_n$ was observed. The advantages of this method include the uniform coating of the nanoparticles or nano film. However, this process has limitations including the higher temperatures required, and it is difficult to scaleup.

## Thermal Decomposition and Pulsed Laser Ablation

Pure and doped metal nanomaterials can be synthesized via decomposing metal alkoxides and salts by applying high energy using heat or electricity. However, the properties of the produced nanomaterials strongly depend on the precursor concentrations, the flow rate of the precursors and the environment. Kim et al. synthesized $TiO_2$ nanoparticles with a diameter less than 30 nm via the thermal decomposition of titanium alkoxide or $TiCl_4$ at 1200°C. Liang et al produced $TiO_2$ nanoparticles with a diameter ranging from 3 to 8 nm by pulsed laser ablation of a titanium target immersed in an aqueous solution of surfactant or deionized water. Nagaveni et al prepared W, V, Ce, Zr, Fe, and Cu iondoped anatase $TiO_2$ nanoparticles by a solution combustion method and found that the solid solution formation was limited to a narrow range of concentrations of the dopant ions. However, the drawbacks of these methods are high cost and low yield, and difficulty in controlling the morphology of the synthesized nanomaterials.

## Templating

The synthesis of nanostructure materials using the template method has become extremely popular during the last decade. In order to construct materials with a similar morphology of known characterized materials (templates); this method utilizes the morphological properties with reactive deposition or dissolution. Therefore, it is possible to prepare numerous new materials with a regular and controlled morphology on the nano and microscale by simply adjusting the morphology of the template material. A variety of templates have been studied for synthesizing titania nanomaterials. This method has some disadvantages including the complicated synthetic procedures and, in most cases, templates need to be removed, normally by calcinations, leading to an increase in the cost of the materials and the possibility of contamination.

## Combustion

Combustion synthesis leads to highly crystalline particles with large surface areas. The process involves a rapid heating of a solution containing redox groups. During combustion, the temperature reaches approximately 650°C for one or two minutes making the material crystalline. Since the time is so short, the transition from anatase to rutile is inhibited.

## Gas Phase Methods

Gas phase methods are ideal for the production of thin films. Gas phase can be carried out chemically or physically. Chemical Vapour Deposition (CVD) is a widely used industrial technique that can coat large areas in a short space of time. During the procedure, titanium dioxide is formed from a chemical reaction or decomposition of a precursor in the gas phase. Physical vapour deposition (PVD) is another thin film deposition technique. Films are formed from the gas phase but without a chemical transition from precursor to product. For $TiO_2$ thin films, a focused beam of electrons heats the titanium dioxide material. The electrons are produced from a tungsten wire heated by a current. This is known as Electron beam (E-beam) evaporation. Titanium dioxide films deposited with E-beam evaporation have superior characteristics over CVD grown films such as, smoothness, conductivity, presence of contaminations and crystallinity. Reduced $TiO_2$ powder (heated at 900°C in a hydrogen atmosphere) is necessary for the required conductance needed to focus an electron beam on the $TiO_2$.

## Microwave Synthesis

Various $TiO_2$ materials have been synthesised using microwave radiation. Microwave techniques eliminate the use of high temperature calcination for extended periods of time and allow for fast, reproducible synthesis of crystalline $TiO_2$ nanomaterials. Corradi prepared colloidal $TiO_2$ nanoparticle suspensions within 5 minutes using microwave radiation. High quality rutile rods were developed combining hydrothermal and microwave synthesis, while $TiO_2$ hollow, open ended nanotubes were synthesised through reacting anatase and rutile crystals in NaOH solution.

## Conventional Sol-gel Method

The sol-gel method is a versatile process used for synthesizing various oxide materials. This synthetic method generally allows control of the texture, the chemical, and the morphological properties of the solid. This method also has several advantages over other methods, such as allowing impregnation or coprecipitation, which can be used to introduce dopants. The major advantages of the sol-gel technique includes molecular scale mixing, high purity of the precursors, and homogeneity of the sol- gel products with a high purity of physical, morphological, and chemical properties. In a typical sol-gel process, a colloidal suspension, or a sol, is formed from the hydrolysis and polymerization reactions of the precursors, which are usually inorganic metal salts or metal organic compounds such as metal alkoxides. A general flowchart for a complete sol-gel process is shown in figure below.

Any factor that affects either or both of these reactions is likely to impact the properties of the gel. These factors, generally referred to as sol-gel parameters, includes type of precursor, type of solvent, water content, acid or base content, precursor concentration, and temperature. These parameters affect the structure of the initial gel and, in turn, the properties of the material at all subsequent processing steps.

After gelation, the wet gel can be optionally aged in its mother liquor, or in another solvent, and washed. The time between the formation of a gel and its drying, known as aging, is also an important parameter. A gel is not static during aging but can continue to undergo hydrolysis and condensation.

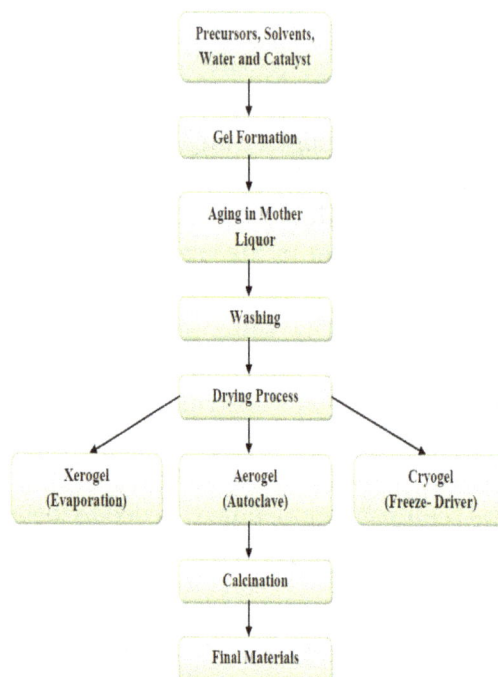

Sol-Gel and Drying Flowchart

Important Parameters in the Various Steps of a Sol- Gel Process.

| Step | Purpose | Important Parameters |
|------|---------|---------------------|
| Solution Chemistry | To form gel | Type of Precursor, Type of Solvent, Water Content, Precursor Concentration, Temperature, pH |
| Aging | To allow a gel undergo changes in properties | Time, Temperature, Composition of the pore liquid, Aging environment |
| Drying | To remove solvent from a gel | Drying method (evaporative super critical & freeze drying),Temperature and heating rate, Pressure and pressurization rate, Time |
| Calcination | To change the physical/chemical properties of the solid, often resulting in crystallization and densification | Temperature and heating rate, Time, Gaseous environment (inert, reactive gases) |

Furthermore, syneresis, which is the expulsion of solvent due to gel shrinkage, and coarsening, which is the dissolution and reprecipitation of particles, can occur. These phenomena can affect both the chemical and structural properties of the gel after its initial formation. Then it must be dried to remove the solvent. Table showed a summary of the key steps in a sol-gel process which includes the aim of each step along with experimental parameters that can be manipulated.

## Aerogel

One important parameter that affects a sol-gel product is the drying condition. Due to the surface tension of the liquid, a capillary pressure gradient is present in the pore walls and this may be able to collapse most of the pore volume when solvent is removed. One convenient way to avoid pore collapse is to remove the liquid from the pores above the critical temperature (Tc) and critical pressure (Pc) of the fluid, namely, supercritical drying. Under supercritical conditions, there is no longer a distinction between the liquid and vapor phases: the densities become equal; there is no liquid-vapor interface and no capillary pressure. This type of drying prevents the formation a liquid-vapor meniscus which recedes during the emptying of the pores in the wet gels. The resulting dried gel, called an aerogel, has a pore volume similar to that of the wet gel.

## Xerogel

Conventional evaporative drying induces capillary pressure associated with the liquid vapor interface within a pore, causing shrinkage of the gel network. In a sample with a distribution of pore sizes, the resultant differential capillary pressure often collapses the porous network during drying. The dried sample often has low surface area and pore volume.

## Cryogel

Another way of avoiding the presence of liquid-vapor interface is to freeze the pore liquid and sublime the resulting solid under vacuum. In this method, the gel liquid is first frozen and thereafter dried by sublimation. Therefore, the formation of a liquid-vapor meniscus is prevented. The materials obtained are then also termed cryogels. Their surface area and mesopore volume tend to be smaller than those of aerogels, although they remain significant. However, freeze-drying does not permit the preparation of monolithic gels. The reason is that the growing crystals reject the gel network, pushing it out of the way until it is stretched to the breaking point. It is this phenomenon that allows gels to be used as hosts for crystal growth: the gel is so effectively excluded that crystals nucleated in the pore liquid are not contaminated with the gel phase; the crystals can grow up to a size of a few millimetres before the strain is so great that macroscopic fractures appear in the gel. Nevertheless, the gel network may eventually be destroyed by the nucleation and growth of solvent crystals, which tend to produce very large pores. To attenuate this event, a rapid freeze process known as flash freezing has been developed. It is also important that the solvent has a low expansion coefficient and a high pressure of sublimation.

## Applications of sol-gel Method

Applications for sol-gel process derive from the various special shapes obtained directly from the gel state (monoliths, films, fibers, and monosized powders) combined with compositional and microstructural control and low processing temperatures. Compared with other methods, such as the solid-state method, the advantages of using sol-gel process include.

- The use of synthetic chemicals rather than minerals enables high purity materials to be synthesized.

- It involves the use of liquid solutions as mixtures of raw materials. Since the mixing is with low viscosity liquids, homogenization can be achieved at a molecular level in a short time.

- Since the precursors are well mixed in the solutions, they are likely to be equally well-mixed at the molecular level when the gel is formed; thus on heating the gel, chemical reaction will be easy and at a low temperature.

- Changing physical characteristics such as pore size distribution and pore volume can be achieved.

- Incorporating multiple components in a single step can be achieved.

- Producing different physical forms of samples is manageable.

## Experimental Procedures and Characterization Techniques

## Chemicals Used

Most of the chemicals used in the research are standard chemicals that are normally available in the laboratory. Special materials for DSSC are mostly bought from Solaronix. Table below shows the list of materials used in this research.

## List of Materials Used

| Chemicals | Manufacturer | Purity | Usage |
|---|---|---|---|
| Titanium (IV)isopropoxide | Sigma-Aldrich | 98% | Nanoparticles |
| Ethanol | Changshu Yangyuan Chemical | 99.9% | Solvent |

| | | | |
|---|---|---|---|
| Acetic acid | Merck | 99 % | Catalysts |
| DI-water | - | - | Hydrolysis |
| Acetone | Merk | 97 % | Clean substrate |
| Methanol | Molychem | 99.8% | Clean substrate |
| Isopropanol | Fisher Scientific | 99 % | Clean substrate |
| Acetylacetone | Sigma - Aldrich | 39.5% | Binder |
| FTO-glass | Solaronix, Spektron | - | Substrate |
| DMF Solvent | Merck | 98 % | Solvent |
| Triton X-100 | Sigma-Aldrich | - | Surfactant |
| Polyethylene glycol(600) | Merck | - | Surfactant |
| N719 dye | Solaronix | - | Sensitizer |
| Rhodamine dye | Sigma - Aldrich | - | Sensitizer |
| Coumarin 30 B | Sigma - Aldrich | - | Sensitizer |
| Plasitol | Solaronix | - | Counter electrode |
| Idolyte TG 50 | Solaronix | - | electrolyte |
| Aluminium nitrate | Loba Chemie | 98 % | Dopant Material |
| Silver nitrate | Qualigens | 99.9 % | Dopant Material |
| Magnesium nitrate | Loba Chemie | 98 % | Dopant Material |
| Nickel nitrate | Merck | 97 % | Dopant Material |
| Chromium nitrate | Himedia | 98 % | Dopant Material |

## Synthesis of Titanium Dioxide Nanopowders

The nano-$TiO_2$ powder was prepared with titanium isopropoxide solution as the raw material. In a typical experiment, 60 ml of deionized water and 5 ml of glacial acetic acid were dissolved at room temperature to obtain solution A. 14 ml titanium isopropoxide was dissolved in 40 ml of anhydrous ethanol with constant stirring to form solution B. Then, the solution B was added drop-wise into the solution A within 60 min under vigorous stirring. Subsequently, the obtained sol was stirred continuously for 2 h and aged for 48 h at room temperature. As-prepared $TiO_2$ gels were dried for 10h at 80°C. The obtained solids were ground and finally calcinated at 450°C for 2 h (heating rate = 3°C/min).

## Substrate Cleaning

Coated glass with highly F-doped Transparent Conducting Oxide (TCO) usually serves as a support for the dye-sensitized oxide. It allows light transmission while providing good conductivity for current collection. The substrates are first dipped into Acetone with ultrasonic bath for 15 minutes to dissolve unwanted organic materials and to remove dust and contamination material that are left on the substrates post manufacture. Another 15 minutes of ultrasonic bath in methanol is followed in order to remove the acetone and materials that are not cleansed or dissolved by acetone. Finally, a 10 minute ultrasonic bath in isoproponal was needed to further remove the residual particles on the substrates.

## $TiO_2$ Photoanode Deposition on FTO

It is very important to work with a fingerprint free Transparent Conducting Oxide (TCO), always gloves were used and TCO was cleaned with alcohol prior to use. TCO was heated to 50 °C at the beginning of the process to increase the adhesion and Scotch 3M adhesive tape were applied on the edges of the conductive side of the TCO glass. The reason for applying tapes was preparing a mould such that nonsintered $TiO_2$ has always same area and thickness for all samples. A certain proportions of nano-$TiO_2$ powder with ethanol, acetylacetone (A.R), polyethylene glycol and triton(X-100) were mixing for 30 minutes in agate mortar. Then $TiO_2$ colloidal was dropped on the conductive side of the TCO after the conductive side of the TCO was checked by the multimeter. Then, the $TiO_2$ paste was uniformly distributed over the TCO by Doctor Blade method. Doctor blade means a film smoothing method using any steel, rubber, plastic, or other type of blade used to apply or remove a liquid substance from another surface.

The term "doctor blade" is derived from the name of a blade used in conjunction with the ductor roll on the letter press. The term "ductor blade" eventually mutates into the term "doctor blade".

## Heat Treatment for Photoanode

The tapes were removed from the glass and plates were sintered at 450 °C for 30 min in air above the $TiO_2$ coated TCO was required after $TiO_2$ material deposition. Colour of $TiO_2$ becomes brown in the middle of the sintering process and then its colour changes to the brownish-white. This colour remained till the end of the sintering process. This is to ensure that the polymer or macromolecules in $TiO_2$ colloid such as Acetylacetone can be removed, leaving tiny holes in $TiO_2$ layers, resulting in better dye absorption and better contact between $TiO_2$ particles. In consequence, it optimizes the chances of electrons being excited by the photons and increases the amount of excited electrons entering into the $TiO_2$ conduction band.

## Preparation of Dye Solution

The dye solution used in this research was N719, Rhodamine B and Coumarin 30B dyes are commonly used in DSSC laboratories. The material is normally available in powder form from commercial companies and dissolved in chemicals before use. The samples need to be fully covered with dye solution for 24 hours for the dye particles to be fully absorbed. Due to the fact that dyes are the light absorbing material, it needs to be store in the dark, preventing the loss of functionality and the samples need to be rinsed with different solvent medium.

Heat Treatment of FTO Coated TiO2 (a) Before and (b) After.

## 3Preparation of Counter Electrode

The platinization procedure given by Solaronix was applied because the material was taken from Solaronix. Actually, this method is simply called thermal decomposition which is most widely used platinization procedure. Plasitol was applied on the surface by using a brush. All TCO glasses were sintered at 400 for 5 minutes for decomposition which was the minimum required calcination condition according to the procedure.

## Nanocrystalline Dye Sensitized Solar Cell Assembly

Sensitized $TiO_2$ photo-anode and the counter electrode were stacked together face to face and the liquid electrolyte, Idolyte TG 50 solution drop penetrated into the working space and counter electrode via capillary action. The two electrodes were held with binder clips. The flow chart of preparation of dye sensitized solar cell and schematic diagram was shown in figures below.

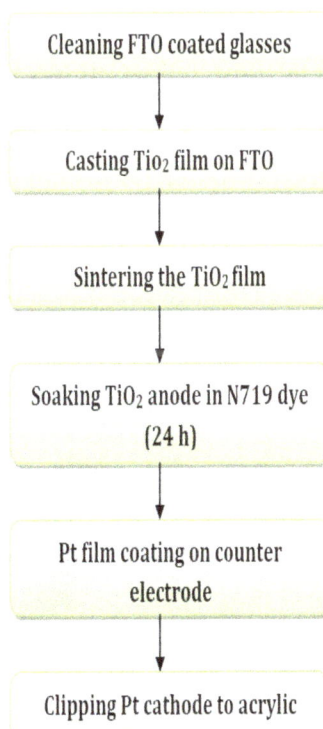

Figure: Preparation of Dye Sensitized Solar Cell

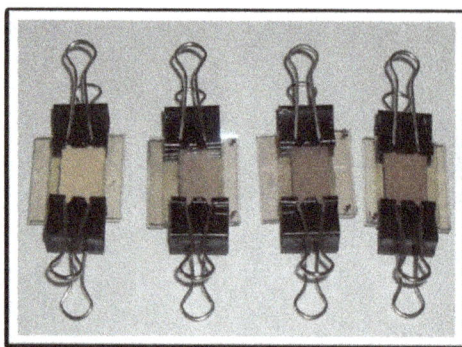

Schematic Diagram of Dye Sensitized Solar Cells.

## Characterization

The synthesized nanomaterials are characterized by the following analytical tools which are described in detailed:

- XRD analysis

- UV-Vis spectroscopy

- Field Emission Scanning Electron Microscopy

- Energy Dispersive X-Ray Spectroscopy

- Photoluminescence spectroscopy

## XRD Analysis

Solid materials are formed by atoms or atomic group arranged in certain way. When an x-ray beam is injected into the material, it would be scattered by atoms. If two or more x-ray beams scattered by the atoms that have some phase differences are superimposed onto each other, diffraction is occurred. The x-ray diffraction instrument (SHIMADZU-6000 Model) is used to collect the intensities of the scattered signals to get the diffraction pattern of the measured sample. This pattern is normally as the signal intensity versus the phase angle. When such a pattern is used in the crystal surface calibration process for the sample, the material's crystalline structure, such as the orientation and phase angles, can be obtained. Identification of the phases was made with the help of the Joint Committee on Powder Diffraction Standards (JCPDS) files. An advantage for using X- ray diffraction measurement is that it can analyse the material without causing damages on the material.

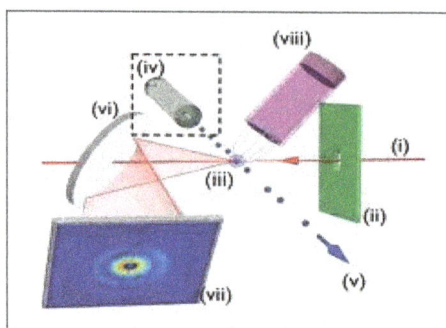

Photograph of a Typical XRD Diffractometer

## UV-Vis Spectroscopy

Many molecules absorb ultraviolet (UV) or visible light. The absorption of UV or visible radiation is caused by the excitation of outer electrons, from their ground state to an excited state. The Bouguer-Lambert-Beer law forms the mathematical physical basis for the light absorption measurements on gas and in solution. According to this law, absorbance is directly proportional to the path length l, and the concentration of the absorbing substance, c, and can be expressed as A = εbc, where ε is a constant of proportionality, called the absorbtivity. In addition, absorption strongly depends on the types of samples, and the environment of the sample. For instance, molecules absorb radiation of various wavelengths depending on the structural groups present within the molecules, and show a number of absorption bands in the absorption spectrum. The solvent in which the absorbing species is dissolved also has an effect on the spectrum of the species. Moreover, the size of the particle is also important. If the size of the particle d>>λ, light interacts with the samples instead of absorption, with parts of the light scattered and reflected.

When dealing with solid samples, light penetrates into the sample; undergoes numerous reflections, refractions and diffraction and emerges finally diffusely at the surface. The Bouguer-Lambert-Beer law cannot handle solid samples, which is based on the assumption that the light intensity is not lost by scattering and reflection processes.

Diffuse reflectance measurements are usually analyzed on the basis of the Kubelka-Munk equation:

$$F\left(R_\infty\right) = \frac{k}{s}\left(1-R\right)^2 / 2\mathrm{R}$$

where k and s are absorption and scattering coefficients respectively, and R is the reflectance at the front face. F(R∞) is termed the Kubelka-Munk function and is proportional to the concentration of the adsorbate molecules. From the onset of the plot of Kubelka-Munk function vs wavelength or photoenergy, the band gap energy of a semiconductor can be easily calculated. However, to measure a diffuse reflectance spectrum, the diffusely reflected light must be collected with an integrated sphere, avoiding secularly reflected light, and using a reference standard ($BaSO_4$ or white standard).

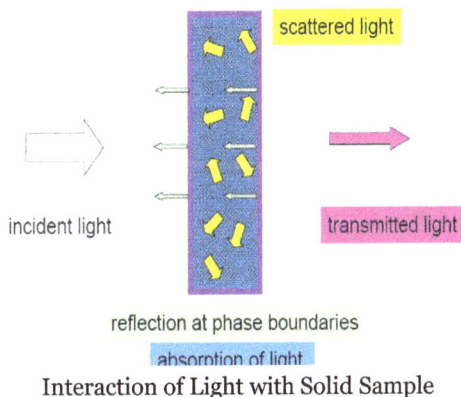

Interaction of Light with Solid Sample

## Field Emission Scanning Electron Microscopy (FESEM)

Electron micrograph images were taken on a FEI QUANTA 200F with a Schottky electron gunas. Measurements were carried out at an accelerating voltage range of 5 – 15 kV. Powdered samples were evenly distributed on a mounted carbon tape surface. Loose powdered sample was removed with canned air spray. The Field Emission Scanning Electron Microscope (FESEM) is a type of electron microscope that images the sample surface by scanning it with a high-energy beam of electrons in a raster scan pattern. The electrons interact with the shells in atoms that make up the sample producing signals that contain information about the sample's surface topography, composition and other properties such as electrical conductivity. The types of signals produced by an SEM include Secondary Electrons (SE), Back Scattered Electrons (BSE), characteristic X-rays, light (cathodoluminescence), Specimen current and Transmitted Electrons (STEM). Generally the most common or standard detection mode is SE imaging. The spot size in a Field Emission SEM is smaller than in conventional SEM and can therefore produce very high-resolution images, revealing details in the range of 1 to 5 nm in size.

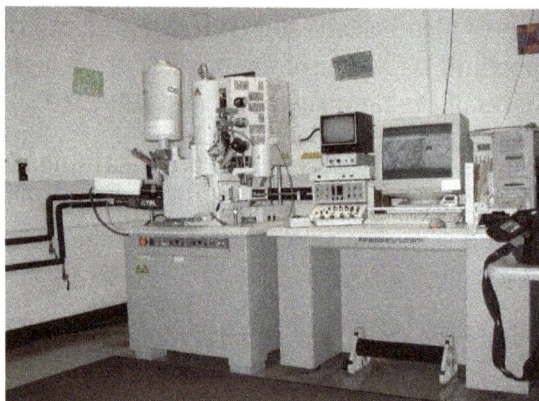

FEI QUANTA 200F

## Energy Dispersive X-ray Spectroscopy (EDS or EDX)

EDS or EDX is an analytical technique used for the elemental analysis or chemical characterization of a sample. It is one of the variants of X-ray fluorescence spectroscopy which relies on the investigation of a sample through interactions between electromagnetic radiation and matter, analyzing X-rays emitted by the matter in response to being hit with charged particles. Its characterization capabilities are due in large part to the fundamental principle that each element has a unique atomic structure allowing X-rays that are characteristic of an element's atomic structure to be identified uniquely from one another.

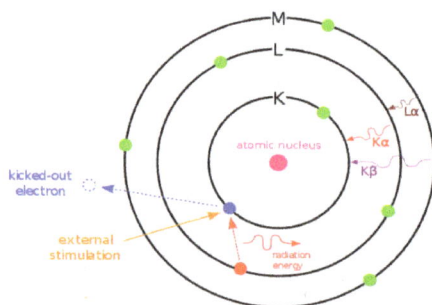

Principle of EDS

The incident beam may excite an electron in an inner shell, ejecting it from the shell while creating an electron hole. An electron from an outer, higher-energy shell then fills the hole, and the difference in energy between the higher-energy shell and the lower energy shell may be released in the form of an X-ray. The number and energy of the X-rays emitted from a specimen can be measured by an energy dispersive spectrometer. As the energy of the X-ray are characteristic of the difference in energy between the two shells, and of the atomic structure of the element from which they were emitted, this allows the elemental composition of the specimen to be measured as shown in the figure above.

## Photoluminescence Spectroscope

Photoluminescence (PL) is the spontaneous emission of light from a material under optical excitation. PL measurement is a kind of powerful and nondestructive technique, which has been carried out on most of semiconductors. To date, there are many different type lasers have been widely used in the PL setup, for example, He-Cd laser with 325 nm, Ar+ laser with 316 nm/514 nm/488 nm, Nd:YAG pulsed laser with 266 nm, tunable solid state lasers and so on.

When we use pump laser to provide pulsed excitation, the lifetime information of excited state can be obtained. Then the setup will be called Time-Resolved PL (TRPL). When light of sufficient energy is illuminated a material, photons are absorbed and excitations are created. These excited carriers relax and emit a photon. Then PL spectrum can be collected and analyzed. However, only the energy of photons is equal to or higher than the bandgap, the absorption can happen in materials.

Therefore, we have to choose different excitation source to do the measurements according to different material with different electronic band structure. The PL peak positions reveal transition energies and the PL intensity implicates the relative rates of radiative and non-radiative recombination. We also can change other external parameters during the PL measurement, such as temperature, excitation power and applied external perturbation such as magnetic field and/or electrical field and/or pressure, which can help us further understand the electronic states and bands.

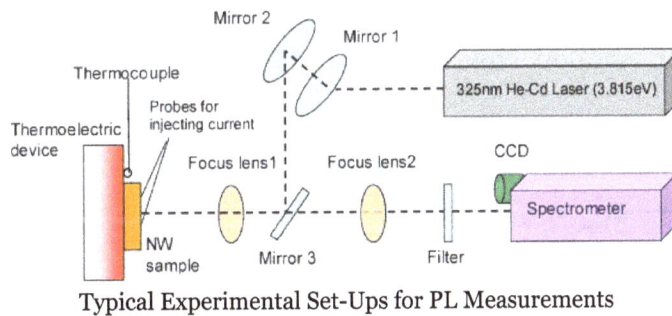

Typical Experimental Set-Ups for PL Measurements

## Solar Cell Efficiency Analysis

The solar cell efficiency is determined by its current-voltage (JV) characteristics under standard illumination conditions. A standard solar spectrum of air mass 1.5 (AM 1.5) with an intensity of 100 W/m2 also referred to as 1 sun, is used for solar cell characterization. The AM 1.5 spectrum corresponds to sunlight that has path through the atmosphere 1.5 times longer than when the sun is directly overhead. The sunlight will be attenuated differently by the earth atmosphere dependent on the incident radiation angle. In the lab, the illumination conditions are provided by a calibrated lamp source. The current-voltage characteristics are monitored under illumination by varying an external load from zero load (short-circuit condition) to infinite load (open-circuit condition). The cell is placed on the simulator stage as shown in figure, where the positive plug is connected to the Pt side and the negative probe is connected to the other substrate with $TiO_2$.

Cell Placement on Solar Simulator

## Mechanical Attrition

Mechanical attrition produces its nanostructures not by cluster assembly but by the structural decomposition of coarser grained structures as a result of plastic deformation. Elemental powders of Al and β-SiC were prepared in a high energy ball mill. More recently, ceramic/ceramic nano-composite WC-14% MgO material has been fabricated. The ball milling and rod milling techniques belong to the mechanical alloying process which has received much attention as a powerful tool for the fabrication of several advanced materials. Mechanical alloying is a unique process, which can be carried out at room temperature. The process can be performed on both high energy mills, centrifugal type mill and vibratory type mill, and low energy tumbling mill.

Different types of milling equipment are available for mechanical alloying and nanoparticle formation.

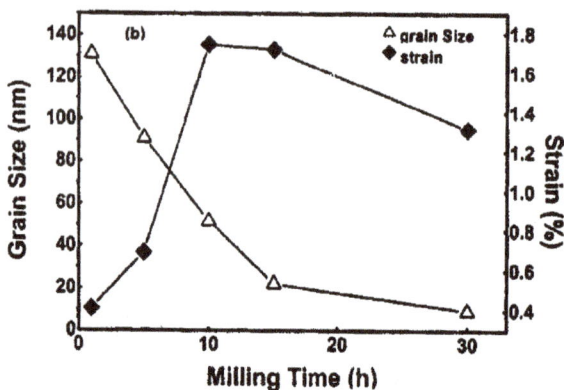

Changes of average grain size and strain for MA Fe-20at.% Co powders with milling time by cyclic operation.

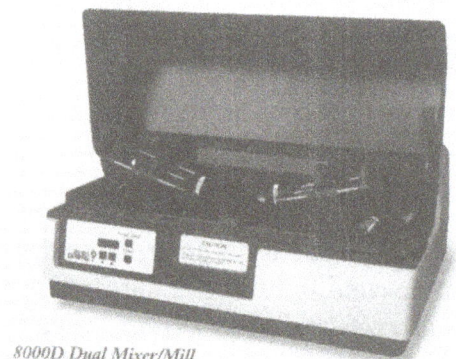

SPEX 8000D dual mixer/mill.

They differ in their capacity, efficiency of milling, and additional arrangements for heat transfer and particle removal. A brief description of the different mills avail- able for MA can be found below.

## SPEX Shaker Mills

Shaker mills such as SPEX, which mill about 10–20 g of powder at a time, are most commonly used for laboratory investigations and for alloy screening pur- poses. The common variety of the mill has one vial, containing the sample and the grinding media, which is secured in the clamp and swung energetically back and forth several thousands times a minute. The back-and-forth shaking motion is combined with lateral movements of the ends of the vial. With each swing of the vial the milling media, typically hard, spherical objects called "milling balls," impact against each other and the end of the vial, both milling and mixing at the same time. Because of the amplitude (about 5 cm) and speed (about 1200 rpm) of the clamp motion, the ball velocities are high (on the order of 5 m/s), and consequently the force of the ball's impact is usually great. Therefore, these mills can be considered a "high-energy" variety.

The most recent design of this mill has provision for simultaneously milling the powders in two vi- als to increase the throughput. This machine incorporates forced cooling to permit extended mill- ing times. A vari- ety of vial materials are available for the SPEX mills, including hardened steel, alumina, tungsten carbide, zirconia, stainless steel, silicon nitride, agate, plastic, and polymethyl- methacrylate. A majority of the research on the fundamental aspects of MA has been carried out with some version of these SPEX mills.

SPEX stainless steel vial set for SPEX 8000D mill

## Planetary Ball Mills

Another popular mill for conducting MA experiments is the planetary ball mill (referred to as Pulverisette) in which a few hundred grams of the powder can be milled at a time. The planetary ball mill owes its name to the planet-like movement of its vials. These are arranged on a rotating support disk, and a special drive mechanism causes them to rotate around their own axes. The cen- trifugal force produced by the vials rotating around their own axes and that produced by the rotating support disk both act on the vial contents, consisting of material to be ground and the grinding balls. Since the vials and the sup- porting disk rotate in opposite directions, the centrifu- gal forces alternately act in like and opposite directions. This causes the grinding balls to run down the inside of the vial—the friction effect—followed by the material being ground. Grinding balls lift off and travel freely through the inner chamber of the vial and collide against the opposing inside wall—the impact effect.

Even though the diskand the vial rotation speeds could not be independently controlled in the earlier versions of this device, it is possible to do so in modern versions. In a single mill there can be either two (Pulverisette 5 or 7) or four (Pulverisette 5) milling stations. Recently, a single- version mill was also developed (Pulverisette 6). Grinding vials and balls are available in a variety of different materials, including agate, silicon nitride, sintered corundum, zirconia, chrome steel, Cr-Ni steel, tungsten carbide, and polyamide. An example of the particles that result from attrition in a planetary mill is shown in figure below.

## Attritor Mills

A conventional ball mill consists of a rotating horizontal drum half-filled with steel balls that range from 0.318 to 0.635 cm in diameter. As the drum rotates the balls drop on the metal powder that is being ground; the rate of grinding increases with the speed of rotation. At high speeds, however, the centrifugal force acting on the steel balls exceeds the force of gravity, and the balls are pinned to the wall of the drum. At this point the grinding action stops. An attritor (a ball mill capable of generating higher energies) consists of a vertical drum with a series of impellers inside it. Set progressively at right angles to each other, the impellers energize the ball charge, causing powder size reduction due to the impact between balls; between the balls and the container wall; and between the balls, the agitator shaft, and the impellers. Some size reduction appears to take place by interparticle collisions and ball sliding. A motor rotates the impellers, which in turn agitate the steel balls in the drum.

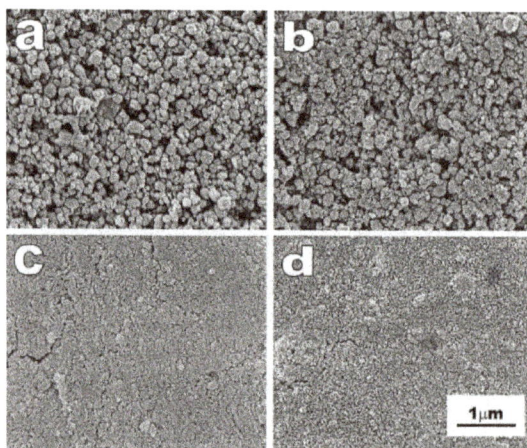

SEM of $Bi_4 Ti_3 O_{12}$ milled for different times: (a) 3, (b) 9, (c) 15, and (d) 20 h.

Attritors are the mills in which large quantities of powder (from about 0.5 to 40 kg) can be milled at a time. Attritors of different sizes and capacities are available. The grinding tanks or containers are available in stainless steel or stainless steel coated with alumina, silicon carbide, silicon nitride, zirconia, rubber, or polyurethane. A variety of grinding media are also available: glass, flint stones, stealite ceramic, mullite, silicon carbide, silicon nitride, Sialon, alumina, zirconium silicate, zirconia, stainless steel, carbon steel, chrome steel, and tungsten carbide.

The operation of an attritor is simple. The powder to be milled is placed in a stationary tank with the grinding media. The mixture is then agitated by a shaft with arms, rotating at a high speed of about 250 rpm. This causes the media to exert both shearing and impact forces on the material. The laboratory attritor works up to 10 times faster than conventional ball mills.

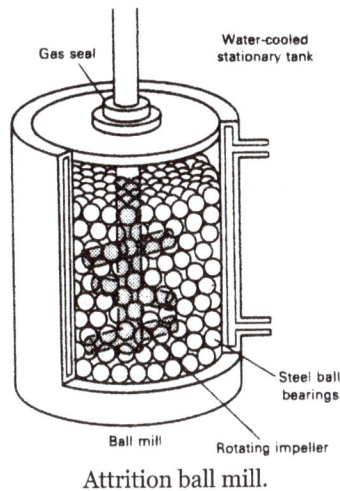

Attrition ball mill.

# Bottoms up Methods

Bottom-up, or self-assembly, approaches to nanofabrication use chemical or physical forces operating at the nanoscale to assemble basic units into larger structures. As component size decreases in nanofabrication, bottom-up approaches provide an increasingly important complement to top-down techniques. Inspiration for bottom-up approaches comes from biological systems, where nature has harnessed chemical forces to create essentially all the structures needed by life. Researchers hope to replicate nature's ability to produce small clusters of specific atoms, which can then self-assemble into more-elaborate structures.

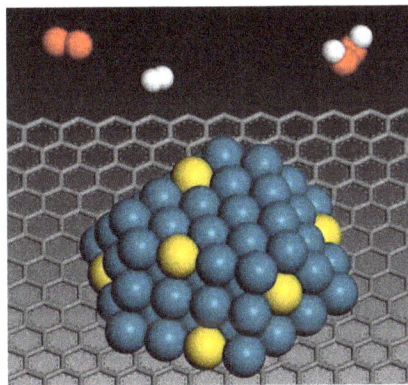

A number of bottom-up approaches have been developed for producing nanoparticles, ranging from condensation of atomic vapours on surfaces to coalescence of atoms in liquids. For example, liquid-phase techniques based on inverse micelles (globules of lipid molecules floating in a nonaqueous solution in which their polar, or hydrophilic, ends point inward to form a hollow core, as shown in the figure) have been developed to produce size-selected nanoparticles of semiconductor, magnetic, and other materials. An example of self-assembly that achieves a limited degree of control over both formation and organization is the growth of quantum dots. Indium gallium arsenide (InGaAs) dots can be formed by growing thin layers of InGaAs on GaAs in such a manner that repulsive forces caused by compressive strain in the InGaAs layer results in the formation of isolated quantum dots. After the growth of multiple layer pairs, a fairly uniform spacing of the dots can be achieved. Another example of self-assembly of an intricate structure is the formation of carbon nanotubes under the right set of chemical and temperature conditions.

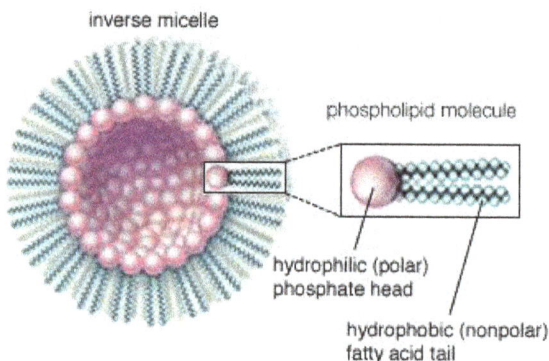

inverse micelle

phospholipid molecule

hydrophilic (polar) phosphate head

hydrophobic (nonpolar) fatty acid tail

DNA-assisted assembly may provide a method to integrate hybrid heterogeneous parts into a single device. Biology does this very well, combining self-assembly and self-organization in fluidic environments where weaker electrochemical forces play a significant role. By using DNA-like recognition, molecules on surfaces may be able to direct attachments between objects in fluids. In this approach, polymers made with complementary DNA strands would be used as intelligent "adhesive tape," attaching between polymers only when the right pairing is present. Such assembly might be combined with electrical fields to assist in locating the attachment sites and then be followed by more-permanent attachment approaches, such as electrodeposition and metallization. There are several advantages of DNA-assisted approaches: DNA molecules can be sequenced and replicated in large quantities, DNA sequences act as codes that can be used to recognize complementary DNA strands, hybridized DNA strands form strong bonds to their complementary sequence, and DNA strands can be attached to different devices as labels. These properties are being explored for ways to self-assemble molecules into nanoscale units. For example, sequences of DNA have been fabricated that adhere only to particular crystal faces of compound semiconductors, providing a basis for self-assembly. By having the correct complementary sequences at the other end of the DNA molecule, certain faces of small semiconductor building blocks can be made that adhere to or repel each other. For example, thiol groups at the end of molecules cause them to attach to gold surfaces, while carboxyl groups can be used for attachment to silica surfaces. Directed assembly is an increasingly important variation of self-assembly where, in quasi-equilibrium environments, parts are moved mechanically, electrically, or magnetically and are placed precisely where they are intended to go.

## Gas Phase Fabrication

Gas phase nanoparticle preparation methods have attracted huge interest over the years due to number of benefits that they can deliver over other methods. These techniques are typically characterized by the ability to accurately control the process parameters to be able to tune shape, size and chemical composition of the nanostructures.

Although, means and methods can differ, almost all gas phase nanomaterial production methods follows following sequence:

• Suspending the precursor materials in a gas phase.

- Transforming the precursor material to small clusters.

- Enforcing the growth of these clusters to a nanoparticles.

- Method to collect prepared nanoparticles.

## Furnace Flow Reactors

Oven sources are the simplest systems to produce a saturated vapor for substances having a large vapor pressure at intermediate temperatures up to about 1700°C. A crucible containing the source material is placed in a heated flow of inert carrier gas. This has the disadvantage that the operating temperature is limited by the choice of crucible material and that impurities from the crucible might be incorporated in the nanoparticles. Nanoparticles are formed by subsequent cooling, such as natural cooling or dilution cooling. For very small particles a rapid temperature decrease is needed which can be achieved by the free jet expansion method described later. Another method is cooling in a turbulent jet. Materials with too low vapor pressure for obtaining appreciable particle density have to be fed in the form of suitable precursors, such as organometallics or metal carbonyls, in the furnace. These decompose in the oven to yield a condensable material. Using furnace flow reactors, elemental nanoparticles such as Ag and Ga and also compounds such as PbS can be obtained.

## Laser Reactors

In the laser pyrolysis technique, being a special class of laser processing techniques, a flowing reactant gas is heated rapidly with an IR laser such as a cw $CO_2$ laser. The source molecules are heated selectively by absorption of the laser beam energy, whereas the carrier gas is only indirectly heated by collisions with the reactant molecules. A gas phase decomposition of the reactants takes place due to the temperature increase and supersaturation is created. As an example, $SiH_4$ pyrolysis results in Si nanoparticle formation and $Fe(CO)_5$ decomposition leads to Fe nanoparticles. The main advantage of laser-heating in gas-flow systems is the absence of heated walls which reduces the danger of product contamination.

UV lasers can be used to obtain photochemical dissociation. Tamir and Berger showed that SiH4 can be dissociated by a two-photon absorption of an ArF excimer laser beam with a wavelength of 193 nm, forming nanoparticles.

## Laser Vaporization of Solids

This technique uses a laser which evaporates a sample target in an inert gas flow reactor. The source material is locally heated to a high temperature enabling thus vaporization. The vapor is cooled by collisions with the inert gas molecules and the resulting supersaturation induces nanoparticle formation. Kato used a continuous-wave $CO_2$ laser with a power of 100 W. Nanoparticles between 6 and 100 nm of many complex refractory oxides such as $Fe_3O_4$, $CaTiO_3$ and $Mg_2SiO_4$ were synthesized under an inert gas pressure of 1 mbar to 5 bar from powders, single crystals or sintered blocks. Nanocomposites can also be produced, as shown e.g. by Chow. They evaporated simultaneously a metallic and a ceramic target by a cw $CO_2$ laser creating a composite film close to the target. The matrix consisted of Al and the dispersed phase consisted of amorphous $SiO_2$ fibers with diameters between 25 and 120 nm.

## Flame Reactors

Nanoparticles are produced by employing the flame heat to initiate chemical reactions producing condensable monomers. The flame route has the advantage of being an inexpensive method, however usually it yields agglomerated particles. An example is the oxidation of $TiCl_4$, $SiCl_4$ or $SnCl_4$ in a $CH_4/O_2$ flame leading to $TiO_2$, $SiO_2$ or $SnO_2$ particles with primary particle sizes between 10 and 100 nm . More complex products can be also obtained. Zachariah et al. added small amounts of $Fe(CO)_5$ and $SiO_2(CH3)_6$ to a premixed $CH_4/O_2$ flame with temperatures at about 2000°C, forming $Fe_3O_4$ particles of 5-10 nm embedded in larger $SiO_2$ host particles. The synthesis was conducted at relatively high temperatures under oxygen-rich conditions to minimize carbon contamination.

A method to reduce the agglomeration of nanoparticles, avoiding oxidation, is to encapsulate them within a material which can be later removed. Dufoux and Axelbaum applied a Na coflow flame with a burner consisting of several concentric tubes. A TiCl4 vapor was fed to the central tube while the Na vapor was fed to the concentric surrounding tube. An inert gas acts as a sheath gas between these two flows and shields the flame from ambient air. Ti particles with sizes of 10-30 nm were produced and were confined within larger NaCl particles. The NaCl can be removed by washing with water, glycerin or by sublimation.

## Plasma Reactors

Plasma can also deliver the energy necessary to cause evaporation or initiate chemical reactions. The plasma temperatures are in the order of 104 °C, decomposing the reactants into ions and dissociating atoms and radicals. Solid powder feeds can also be decomposed by the plasma. Nanoparticles are formed upon cooling while exiting the plasma region. Main types of the thermal plasmas are dc plasma jet, dc arc plasma and rf induction plasma. A small laboratory plasma system, the inductively coupled plasma (ICP), is often used at atmospheric pressures in combination with an aerosol spray system employing an ultrasonic atomizer. This technique is called spray-ICP. As the residence time of the droplets in the plasma is very short it is important that the droplet sizes are small in order to obtain complete evaporation. Complex materials such as multicomponent oxides can be obtained by using appropriate mixtures. By using a solution of $Ba(NO_3)_2$ and $Fe(NO_3)_3$ in water, Mizoguchi obtained by spray-ICP $BaFe_{12}O_{19}$ nanoparticles with sizes between 10 and 50 nm. In most cases, the plasma-generated vapors are quenched by mixing with a cold gas. This leads to high cooling rates, but also to nonuniform cooling which deteriorates the uniformity of the produced particles. Rao used a subsonic nozzle expansion after a dc arc plasma in order to obtain more uniform cooling rates.

Another method uses microwave-generated plasma, in which temperatures between 300 °C and 900 °C can be obtained. The plasma enhances the kinetics of the chemical reactions leading to nanoparticle formation due to ionization and dissociation of the reactive molecules. Since the lower temperature decreases the sinter rate, the formation of hard agglomerates is reduced.

## Spark Source and Exploding Wire

A high-current spark between two solid electrodes can be used to evaporate the electrode material for creating nanoparticles. At the electrodes a plasma is formed. This technique is used for materials with a high melting point such as Si or C, which cannot be evaporated in a furnace. Saunders et al. (1993a) used an electric spark with an energy between 50 mJ and 150 mJ to evaporate material from crystalline Si electrodes. A continuous flow of Ar transports Si crystallites formed, of sizes 2 - 4 nm, to the collection substrate. Reactive evaporation is also possible by adding a suitable reactant gas. As an example, SiC nanoparticles were formed when an arc source generated a discharge between two Si electrodes in 500 mbar of $CH_4$, and $Al_2O_3$ was obtained from Al electrodes in a mixture of 130 mbar $O_2$ and 420 mbar Ar.

A closely related technique is the exploding wire, in which the wire material is vaporized instantaneously by a capacitor in a chamber filled with an inert gas. This method has the disadvantage of being non-repetitive. As an example, GaAs nanoparticles were synthesized by using a GaAs wire.

## Sputtering

Sputtering is a method of vaporizing materials from a solid surface by bombardment with high-velocity ions of an inert gas, causing an ejection of atoms and clusters. Sputter sources such as an ion gun or an hollow-cathode plasma sputter source are normally used in vacuum systems, below $10^{-3}$ mbar, as a higher pressure hinders the transportation of the sputtered material. Instead of ions, electrons from an electron gun can be also used. Sputtering has the advantage that it is mainly the target material which is heated and that the composition of the sputtered material is the same as that of the target.

## Inert Gas Condensation

One of the earliest methods used to synthesize nanoparticles, is the evaporation of a material in a cool inert gas, usually He or Ar, at low pressures conditions, of the order of 1 mbar. It is usually called 'inert gas evaporation'. Common vaporization methods are resistive evaporation, laser evaporation and sputtering. A convective flow of inert gas passes over the evaporation source and transports the nanoparticles formed above the evaporative source via thermophoresis towards a substrate with a liquid N2 cooled surface. A modification which consists of a scraper and a collection funnel allows the production of relatively large quantities of nanoparticles, which are agglomerated but do not form hard agglomerates and which can be compacted in the apparatus itself without exposing them to air. Increased pressure or increased molecular weight of the inert gas leads to an increase in the mean particle size. This so-called Inert Gas Condensation method is already used on a A review of synthesis of nanoparticles in the gas phase for electronic, optical and magnetic applications commercial scale for a wide range of materials. Also reactive condensation is possible, usually by adding $O_2$ to the inert gas in order to produce nanosized ceramic particles. Another method replaces the evaporation boat by a hot-wall tubular reactor into which an organometallic precursor in a carrier gas is introduced. This process is known as Chemical Vapor Condensation referring to the chemical reactions taking place as opposed to the inert gas condensation method. Finally, the Gas Deposition Method is also used in industry. In this method, nanoparticles are formed by evaporation in an inert gas at atmospheric pressure and transported by a special designed transfer pipe to the spray chamber at a pressure of about 0.3 mbar. By moving the nozzle at the end of the transfer pipe, the particles which have a mean velocity of 300 m/s can be deposited in required places on the substrate in the spray chamber. Using this technique writing micron-sized patterns was demonstrated.

## Expansion-cooling

Expansion of a condensable gas through a nozzle leads to cooling of the gas and a subsequent homogeneous nucleation and condensation. Turner et al. showed by numerical simulations that by expanding an organometallic precursor in N2 in a subsonic nozzle results in the formation of particles of about 100 nm with a relative narrow size distribution. The expansion caused the pressure to decrease from 2 bar to 0.75 bar. In order to produce nanoparticles smaller than 5 nm, supersonic free jets expanding in a vacuum chamber with pressures smaller than 10-4 mbar have been used. an inert gas containing a metal vapor was subjected to multiple expansions. Nuclei are formed as a result of two sonic expansions. Then a nuclei growth region in a subsonic, low-pressure reactor produced nanoparticles with mean sizes below 2.5 nm. The separation of nucleation and condensation processes results in a narrower size distribution than usually achieved by supersonic expansion. Further growth is stopped by expansion in a vacuum. Converging nozzles which create an adiabatic expansion in a low-pressure flow have also been used to produce nanoparticles. Although the particles sizes are larger than in a vacuum expansion, particles of the order of 100 nm were obtained with a relatively high production rate. Rao produced 4-10 nm sized nanoparticles by expanding a thermal plasma carrying vapor-phase precursors through a ceramic-lined subsonic nozzle. A special nozzle design minimizes boundary layer effects and approaches a one-dimensional temperature gradient in the flow direction. This leads to a highly uniform quench rate and thus to nanoparticles with a narrow size distribution.

## Liquid Phase Fabrication

### Solvothermal Process

A solvothermal process can be defined as "a chemical reaction in a closed system in the presence of a solvent (aqueous and non aqueous solution) at a temperature higher than that of the boiling point of such a solvent". Consequently a solvothermal process involves high pressures. The selected temperature (sub- or supercritical domains) depends on the required reactions for obtaining the target-material through the involved process.

In the case of aqueous solutions as solvent, the hydrothermal technology have been studied and developed a long time ago with different objectives: (i) mineral extraction (as for leaching ores), (ii) investigation of the synthesis of geological materials, (iii) synthesis of novel materials (iv) crystal growth – in particular the elaboration of α-quartz single crystals due to its piezoelectric properties, (v) the deposition of thin films, (vi) the development of sintering processes in mild conditions, (vii) the elaboration of fine particles well defined in size and morphology.

Hydrothermal processes: due in particular to the chemical composition of water as solvent - is mainly appropriated to the preparation of hydroxides, oxihydroxides or oxides versus the temperature value. The development of non-oxide materials (in particular nitrides, chalcogenide) for investigating their physical properties and for industrial applications required the development of new processes involving non aqueous solvents. Consequently, if solvothermal reactions is a "generic term" for a chemical reaction in a close system in presence of a solvent, these reactions are mainly developed with non aqueous solvents for preparing non-oxide materials.

Two types of parameters are involved in solvothermal reactions:

- The chemical parameters,
- The thermodynamical parameters.

## Chemical Parameters

Two different parameters can be taken into account: the nature of the reagents and the nature of the solvent.

The chemical composition of the precursors must be appropriated to that of the targetmaterials. In addition, the concentration of the precursors seems to play a role on the control of the shape of nanocrystallites resulting of a solvothermal process. Q. WANG through the solvothermal preparation of CdSe and CeTe nanocrystals have claimed the control of the crystallites-shape (dot, rod) with the concentration of the precursors. The interactions between reagents and solvent play an important role in the solvothermal reactions.

The selection of the solvent plays a key-role through the control of the chemical mechanisms leading to the target-material.

The reaction mechanisms induce, during the solvothermal reactions, are dependent on the physico-chemical properties of the solvent. For example LI have described the preparation of $CuTe_4$ using $CuCl_2$, $H_2O$ and tellurium as reagents and ethylenediamine as solvent. Using the same experimental conditions but changing only the nature of the solvent (benzene or diethylamine), tellurium did not react with copper chloride. Compare to non-polar solvent as benzene, ethylenediamine is a polarizing solvent such a property being able to increase the solubility of the reagents. In addition its complexing properties can play an important role in the reaction mechanisms.

The complexing properties of the solvent can lead to the intermediate formation of stable complexe systems $(M(en)_3^{2+})$. Such a complex-cation can act as a template due to its octahedral geometry and can be incorporated into the structure of the final material. This type of solvothermal reactions has led to the synthesis of Sb(III) and Sb(V) thioantimonates $[Mn(en)_3]_2Sb_2S_5$ and $[Ni(en)_3(Hen)]SbS_4$.

In some cases the formation of complex-cations is important as an intermediate step during the solvothermal reaction mechanisms. This is the case of the solvothermal preparation of the semiconductor material $CuInSe_2$. The starting products were $CuCl_2$, $InCl_3$ and Se. The solvent was either ethylenediamine (en) or diethylamine. The selected experimental conditions were 180°C and the resulting autogeneous pressure. The propose reaction mechanisms involve four steps:

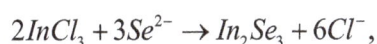

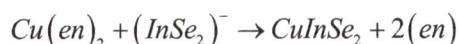

$$2InCl_3 + 3Se^{2-} \rightarrow In_2Se_3 + 6Cl^-,$$

$$In_2Se_3 + Se^{2-} \rightarrow 2(InSe_2)^-,$$

$$Cu^+ + 2en \rightarrow Cu(en)_2^+$$

$$Cu(en)_2 + (InSe_2)^- \rightarrow CuInSe_2 + 2(en)$$

The nucleophilic attack by amine could activate selenium to form $Se^{2-}$ in a similar way that sulphur is activate by amine to $S^{2-}$. The formation of the $Cu(en)^{2+}$ complex ($Cu^+$ resulting from the in situ reduction of $Cu^{2+}$) seems to play are important role in controlling the nucleation and growth of CuInSe2 nano-whiskers. Replacing ethylenediamine by ethylamine as solvent, the reactivity is lowered and the resulting morphology consists on spherical particles of $CuInSe_2$.Consequently the nature of the solvent can act on the reactivity and the morphology of the resulting crystallites.

The physico-chemical properties of the selected solvent can also play an important role for orienting the structural form of the final material. LU have underlined that the solvothermal synthesis of MnS can lead to metastable ($\beta$ and $\gamma$) or stable ($\alpha$) structural forms versus the composition of the solvent. Using $MnCl_2.4H_2O$ and $SC(NH_2)_2$ as reagents and either an hydrothermal reaction (water as solvent) and or a solvothermal reaction (ethylenediamine as solvent), the stable form ($\alpha$-MnS) with the rocksalt structure was observed. With the same reagents but with benzene as solvent, the wurtzite type structure ($\gamma$-MnS) was prepared, with tetrahydrofurane (THF) only the zinc-blende structure ($\beta$-MnS) can be observed.

The stabilization of different structural forms: stable $\alpha$ form or metastable forms ($\beta$, $\gamma$) versus water and the two others solvents (benzene and tetrahydrofurane) can be attributed to the ability to form a stable Mn complex $\left( Mn(H_2O)_6^{2+} \text{ or } Mn(en)_3^{2+} \right)$ during the reaction mechanisms. The difference observe between benzene and THF suggests that a non polar solvent ($C_6H_6$) is more appropriated for stabilizing the wurtzite-form ($\gamma$-MnS). Consequently the solubility of the $Mn^{2+}$ precursor appears to play also an important role for orienting the stabilization of a stable structural form.

Another example is the selective synthesis of $KTaO_3$ either as perovskite or pyrochlore structure versus the composition of the mixed solvents (water-ethanol or water-hexane systems) with a KOH concentration one order of magnitude lower than that in conventional processes.

The oxidation-reduction properties of the solvothermal medium during the reaction can be induced by the nature of the solvent or the composition of mixed solvents and by the use of additives.

The solvothermal processing of $Sb(III)Sb(V)O_4$ nanorods from $Sb_2O_5$ powder involves the reducing properties of ethylenediamine as solvent. At the same temperature (200 °C), if the reaction time is one day only $Sb(III)Sb(V)O_4$ nanorods are formed but after 3 days only metallic Sb particles are observed.

The formation of copper(I) chloride particles with tetrapod-like-morphology used a mixture of acetylacetone and ethylene-glycol as solvent (50/50) and $CuCl_2.2H_2O$ as precursor. During the solvothermal processing of such particles acetylacetone acts as reducing agent ($Cu^{2+} \rightarrow Cu^+$) whereas ethylene-glycol favorizes the anisotropic shape for CuCl crystallites.

On the contrary the solvothermal preparation of InAs as nanoscale semiconductor from $InCl_3$ and $AsCl_3$ as reagents and xylene as solvent requires the use of Zn metal particles as additive. The reaction mechanisms could be described as a co-reduction route:

$$In^{3+} \rightarrow In^0 \text{ and } As^{3+} \rightarrow As^0,$$

through the reaction: $InCl_3 + AsCl_3 + 3Zn \rightarrow InAs + 3ZnCl_2$.

Another interesting illustration of the use of reducing agent in addition of the reagents involves the preparation of the mixed-valent spinel $CuCr_2Se_4$ which is metallic and ferromagnet with a Curie temperature of 450K.

## The Thermodynamical Parameters

These parameters are: temperature, pressure and the reaction time. The solvothermal reactions are mainly developed in mild temperature conditions : (T<400 °C) .Temperature and pressure improving in the major cases the solubility, the increase of such parameters induces an enhancement of the precursors-concentration into the solvent and then favors the growing process (in particular in the preparation of micro- or nanocrystallites).

## Development of Solvothermal Reactions

## Reactions Involved in Solvothermal Processes

Solvothermal reactions involve "in situ" different reaction-types as mentioned through the analysis of the chemical factors governing such processes. In particular, it is possible in a first approach to classify the reactions in approximately 5 types

(i)  oxidation-reduction,

(ii) hydrolysis,

(iii) thermolysis,

(iv) complex-formation,

(v) Metathesis reactions.

The development of these different reactions implies to control carefully the chemistry in non-aqueous solvents and consequently to get more informations concerning the physicochemical properties of such solvents.

## Main Applications of Solvothermal Processes

Solvothermal reactions have been developed in different scientific domains:

- The synthesis of novel materials (design of materials with specific structures and properties).

- The processing of functional materials (an emerging route in synthesis chemistry).

- The crystal growth at low-temperature (a way to single crystals of low-temperature forms or with à low density of defects).

- The preparation of micro- or nanocrystallites well define in size and morphology (as precursors of fine structured ceramics, catalyst, elements of nano-device).

- The low- temperature sintering (preparation of ceramics from metastable structural forms, low temperature forms or amorphous materials).

- The thin films deposition (with the development of low-temperature processes).

## Sol-gel Process

This method is performed in the liquid phase. It is a useful self-assembly process for fabricating nanoparticles as well as nanostructured surfaces and three-dimensional nanostructured materials such as aerogels.

A "sol" is a type of colloid1 in which a dispersed solid phase is mixed in a homogeneous liquid medium. An example of a naturally occurring sol is blood. As the name suggests, the sol-gel process involves the evolution of networks through the formation of a colloidal suspension (sol) and gelation of the sol to form a network in a continuous liquid phase (gel).

The first stage in the sol-gel process is the synthesis of the colloid. The precursors are normally ions of a metal. Metal alkoxides and alkoxysilanes are the most popular since they react readily with water (hydrolysis). The most widely used alkoxysilanes are tetramethoxysilane (TMOS) and tetraethoxysilane (TEOS) which form silica gels. Alkoxides such as aluminates, titanates and borates are also used, often mixed with TMOS or TEOS. In addition, since alkoxides and water are immiscible, a mutual solvent is used, such as an alcohol.

The sol-gel process involves four steps. First the hydrolysis reaction, in which the -OR group is replaced with an -OH group. The hydrolysis reaction can occur without a catalyst but is more rapid and complete when they are used. As in any hydrolysis reaction the catalyst can be a base (NaOH or NH3) or an acid (HF or CH3COOH).

After hydrolysis, the sol starts to condense and polymerise. This leads to a growth of particles which, depending on various conditions such as pH, reach dimensions of a few nanometres. The condensation/polymerisation reaction is quite complex and involves many intermediate products, including cyclic structures. The particles then agglomerate: A network starts to form throughout the liquid medium, resulting in thickening, which forms a gel.

All the four steps illustrated above are affected by the initial conditions for the hydrolysis reaction and the condensation/polymerisation. These conditions include pH, temperature and time of reaction, nature of the catalyst, etc.

The sol-gel process is very commonly used to make silica gels. Other type of gels can also be formed; aluminosilicate gels are special because they form tubular structures. One such a product is imogolite which has an external diameter of about 2.5 nm and internal tube diameter of 1.5nm. These types of nanostructures are known to be good adsorbents for anions such as chloride, chlorate, sulphate and phosphates. The imogolite structure can be dissolved away with HF. Therefore these nanostructures can be used for template synthesis: the tube can be filled with atoms and then dissolved away, leaving rows of atoms (2.5 atoms of gold in a row measure 1 nm).

summarises the different sol-gel processes. To make the most of the large surface area of nanoparticles, the gel can be placed on a surface. This way a greater bulk-area ratio is obtained. Another strategy is to form an aerogel. These are three-dimensional continuous networks of particles with air (or any other gas) trapped at their interstices. Aerogels are characterised by being porous and very light yet able to withstand 100 times their weight.

A versatile way to create ordered surface nanostructures is to perform the sol-gel synthesis in a liquid which is itself ordered. Liquid crystals are precisely this: they have a crystalline structure but exist in a liquid (rather than solid) phase. Nanostructured silica with controlled pore size, shape and ordering can be made in this way.

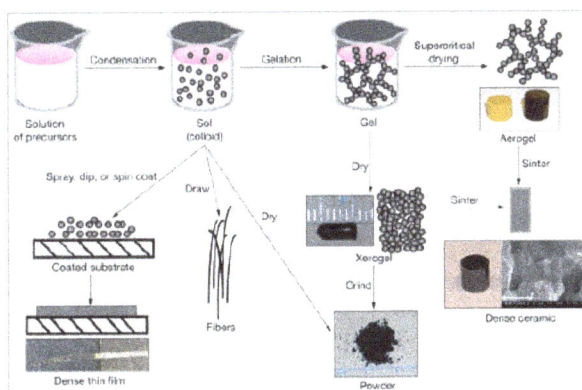

Schematic overview of different materials that can be obtained through a sol-gel process

The liquid crystalline casting method described above can also be used to produce nanostructured metals. This development is very useful for making nanostructured catalytic surfaces, like platinum or palladium surfaces. Since these metals are very rare and expensive, it is highly advantageous to have surfaces where nearly all metal atoms can take part in the catalytic reaction being on the surface.

# Pyrolysis

Pyrolysis is a process of chemically decomposing organic materials at elevated temperatures in the absence of oxygen. The process typically occurs at temperatures above 430°C (800°F) and under pressure. It simultaneously involves the change of physical phase and chemical composition, and is an irreversible process. The word pyrolysis is coined from the Greek words "pyro" which means fire and "lysis" which means separating.

Pyrolysis is commonly used to convert organic materials into a solid residue containing ash and carbon, small quantities of liquid and gases. Extreme pyrolysis, on the other hand yields carbon as the residue and the process is called carbonization. Unlike other high-temperature processes like hydrolysis and combustion, pyrolysis does not involve reaction with water, oxygen or other reagents. However, as it is practically not possible to achieve an oxygen- free environment, a small amount of oxidation always occurs in any pyrolysis system.

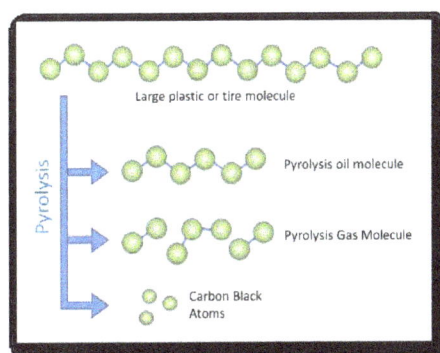

There are three types of pyrolytic reactions differentiated by the processing time and temperature of the biomass.

## Slow Pyrolysis

Slow pyrolysis is characterized by lengthy solids and gas residence times, low temperatures and slow biomass heating rates. In this mode, the heating temperatures ranges from 0.1 to 2 °C (32.18 to 35.6 °F) per second and the prevailing temperatures are nearly 500 °C (932 °F). The residence time of gas may be over five seconds and that of biomass may range from minutes to days. During slow pyrolysis, tar and char are released as main products as the biomass is slowly devolatilized. Repolymerization/recombination reactions occur after the primary reactions take place.

## Flash Pyrolysis

Flash pyrolysis occurs at rapid heating rates and moderate temperatures between 400 and 600 °C (752 and 1112 °F). However, vapor residence time of this process is less than 2s. Flash pyrolysis produces fewer amounts of gas and tar when compared to slow pyrolysis.

## Fast Pyrolysis

This process is primarily used to produce bio-oil and gas. During the process, biomass is rapidly heated to temperatures of 650 to 1000 °C (1202 to 1832 °F) depending on the desired amount of bio- oil or gas products. Char is accumulated in large quantities and has to be removed frequently.

Microwave Pyrolysis Fast pyrolysis has been shown to benefit from the use of microwave heating. Biomass typically absorbs microwave radiation very well, making heating of the material highly efficient - just like microwave heating of food, it can reduce the time taken to initiate the pyrolysis reactions, and also greatly reduces the energy required for the process. Because microwave heating can initiate pyrolysis at much lower overall temperatures (sometimes as low as 200-300 °C), it has been found that the bio-oil produced contains higher concentrations of more thermally labile, higher value chemicals, suggesting that microwave bio-oil could be used as a replacement to crude oil as a feedstock for some chemical processes.

## Advantages of Pyrolysis

### Process

Pyrolysis transforms organic materials into their gaseous components, a solid residue of carbon and ash, and a liquid called pyrolytic oil (or bio-oil). Pyrolysis has two primary methods for removing contaminants from a substance: destruction and removal. In destruction, the organic contaminants are broken down into compounds with lower molecular weight, whereas in the removal process, they are not destroyed but are separated from the contaminated material. Pyrolysis is a useful process for treating organic materials that "crack" or decompose under the presence of heat; examples include polychlorinated biphenyls (PCBs), dioxins, and polycyclic aromatic hydrocarbons (PAHs). Although pyrolysis is not useful for removing or destroying inorganic materials such as metals, it can be used in techniques that render those materials inert.

## Applications

Pyrolysis has numerous applications of interest to green technology. It is useful in extracting materials from goods such as vehicle tires, removing organic contaminants from soils and oily sludges, and creating biofuel from crops and waste products. Pyrolysis can help break down vehicle tires into useful components, thus reducing the environmental burden of discarding the tires. Tires are a significant landfill component in many areas, and they release PAHs and heavy metals into the air when they are burned. However, when tires are pyrolyzed, they break down into gas and oil (usable for fuel) and carbon black (usable as filler in rubber products, including new tires, and as activated charcoal in filters and fuel cells). In addition, pyrolysis can remove organic contaminants, such as synthetic hormones, from sewage sludge (semisolid materials that remain after wastewater is treated and the water content reduced) and make heavy metals remaining in the sludge inert, which allows the sludge to be used safely as fertilizer.

Furthermore, pyrolyzing biomass (biological materials such as wood and sugarcane) holds great promise for producing energy sources that could supplement or replace petroleum-based energy. Pyrolysis causes the cellulose, hemicellulose, and part of the lignin in the biomass to disintegrate to smaller molecules in gaseous form. When cooled, those gases condense to the liquid state and become bio-oil, while the remainder of the original mass (mainly the remaining lignin) is left as solid biochar and noncondensable gases.

## Advantages of Pyrolysis

The key benefits of pyrolysis include the following:

1. It is a simple, inexpensive technology for processing a wide variety of feedstocks.

2. It reduces wastes going to landfill and greenhouse gas emissions.

3. It reduces the risk of water pollution.

4. It has the potential to reduce the country's dependence on imported energy resources by generating energy from domestic resources.

5. Waste management with the help of modern pyrolysis technology is inexpensive than disposal to landfills.

6. Construction of a pyrolysis power plant is a relatively rapid process.

7. It creates several new jobs for low-income people based on the quantities of waste generated in the region, which in turn provides public health benefits through waste cleanup.

## References

- Technology, nanotechnology, Nanofabrication-236454: britannica.com, Retrieved 28 May 2018

- Gas-phase-synthesis: inithi.wordpress.com, Retrieved 07 July 2018

- Derivate-Servlet, Derivate-5148: duepublico.uni-duisburg-essen.de, Retrieved 15 July 2018

- Pyrolysis-Plant, pyrolysis-plant-351: wastetireoil.com, Retrieved 25 April 2018

- Pyrolysis-321388: britannica.com, Retrieved 19 March 2018

# 5

# Types of Nanomaterials

Nanomaterials are classified into groups, namely nanostructured materials and nano-objects. An elaborate study of the varied types of nanomaterials has been provided in this chapter, which includes topics such as carbon based nanomaterials, metal based nanomaterials, dendrimers and composites.

## Carbon-based Nanomaterials

Carbon-based nanomaterials (CBNs), namely, fullerene, carbon nanotubes, and graphene, have attracted significant attention since their discoveries, and in these days they play significant role in nanoscience and nanotechnology. The unique properties of CBNs make them widely used in many fields ranging within material science, energy, environment, biology, medicine, and so forth.

Carbon-based nanomaterials (CBNs) have become important due to their unique combinations of chemical and physical properties (i.e., thermal and electrical conductivity, high mechanical strength, and optical properties), extensive research efforts are being made to utilize these materials for various industrial applications, such as high-strength materials and electronics.

Graphite is one of the oldest and most widely used natural materials. More traditionally known as the main ingredient of pencil lead, from which the name "graphite" originated, it is now more widely used in several large-scale industrial applications, such as carbon raising in steelmaking, battery electrodes, and industrial-grade lubricants Due to its high demand, the consumption of synthetic graphite has significantly increased in recent years. Extensive scientific investigation into graphite has revealed that its unique combination of physical properties stems from its macromolecular structure, which consists of stacked layers of hexagonal arrays of sp carbon.

With the deeper appreciation and development of nanofabrication techniques and nanomaterials that have progressed within the last two decades, graphite is now being actively used as a starting material to engineer various types of carbon-based nanomaterials (CBNs), including single or multi-walled nanotubes, fullerenes, nanodiamonds, and graphene These CBNs possess excellent mechanical strength, electrical and thermal conductivity, and optical properties; much of the research efforts have been focused on utilizing these advantageous properties for various applications, such as high-strength composite materials and electronics.

Carbon-based nanomaterials have been widely regarded as highly attractive biomaterials due to their multi-functional nature. In addition, incorporating CBNs into existing biomaterials could further augment their functions. Therefore, CBNs have found their way into many areas of biomedical research, including drug delivery systems, tissue scaffold reinforcements, and cellular sensors.

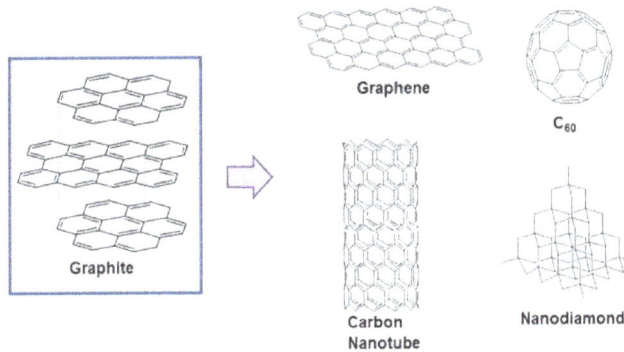

Graphite → Graphene, Carbon Nanotube, C$_{60}$, Nanodiamond

## Carbon Nanotubes

Ever since their discovery, carbon nanotubes (CNTs) have become the most widely used CBNs. Carbon nanotubes are commonly synthesized by arc discharge or chemical vapor deposition of graphite. They have a cylindrical carbon structure, and possess a wide range of electrical and optical properties stemming not only from their extended sp-carbon, but also from their tunable physical properties (e.g., diameter, length, single-walled vs. multi-walled, surface functionalization, and chirality). Due to the diverse array of their useful properties, CNTs have been explored for use in many industrial applications. For example, CNTs are well known for their superb mechanical strength: their measured rigidity and flexibility are greater than that of some commercially available high-strength materials (e.g., high tensile steel, carbon fibers, and Kevlar). Thus, they have been utilized as reinforcing elements for composite materials such as plastics and metal alloys, which have already led to several commercialized products. However, the possibility of CNT-incorporated composites as super high-strength load-bearing materials has not been met with satisfactory results, mostly due to their poor interaction with the surrounding matrices, which leads to inefficient load transfer from the matrices to the CNTs.

Many recent research efforts have been geared toward incorporating CNTs into various materials to utilize their multi-functional nature (i.e., electrical and thermal conductivity, and optical properties) rather than focusing purely on composite mechanical strength. For example, the excellent electrical properties of CNTs coupled with their nano scale dimensions are of great interest in electronics for the construction of nano scale electronic circuitry. In addition, CNTs are known to have low threshold electric fields for field emission, as compared with other common field emitters. Thus, CNTs are actively explored in high-efficiency electron emission devices such as electron microscopes, flat display panels, and gas-discharge tubes. Carbon nanotubes also display strong luminescence from field emission, which could be used in lighting elements.

## Graphene

Graphene is the latest nanomaterial to burst onto the scene. The ground-breaking work by Geim and Novoselov provided a simple method for extracting graphene from graphite via exfoliation and explored its unique electrical properties. Graphene and CNTs possess similar electrical, optical, and thermal properties, but the two-dimensional atomic sheet structure of graphene enables more diverse electronic characteristics; the existence of quantum Hall effect and massless Dirac fermions help explain the low-energy charge excitation at room temperature and the optical transparency in infrared and visible range of the spectrum. In addition, graphene is structurally robust yet highly flexible, which makes it attractive for engineering thin, flexible materials.

### Other Carbon-based Nanomaterials

Buckminsterfullerene, also commonly known as the buckyball, is a spherical closed-cage structure (truncated icosahedron) made of sixty sp carbon. Its discovery in 1985 and subsequent investigation led to the uncovering of electronic properties, stemming from its highly symmetrical structure, and potential applications, it can be argued that the scientific pursuit of CBNs and their potential applications began with the discovery of $C^{60}$. The popularity of $^C60$ has somewhat diminished in recent years with the rise of more scalable and practical CBNs such as CNTs and graphene. However, its uniform size and shape as well as availability for chemical modification led many scientists to develop $C^{60}$ derivatives for therapeutic purposes. Perhaps the most fascinating and highly promising aspect of $^C60$ is its anti-human immunodeficiency virus (HIV) activity. Schinazi first discovered a group of water-soluble $C^{60}$ derivatives capable of inhibiting HIV protease activity by binding to its active site, due to their unique molecular structure and hydrophobicity. Various $^C60$ derivatives have since been developed that display anti-HIV activity by targeting other important HIV enzymes, such as reverse transcriptase. These results demonstrate that $C^{60}$ p derivatives may become a potent group of AIDS therapeutics in the future.

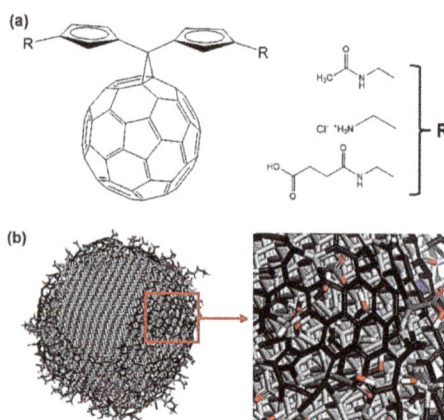

Nanodiamond (ND) has also generated interest in the field of biomedical engineering in recent years. Nanodiamonds are synthesized by high energy treatment of graphite, most commonly via detonation, and are smaller than 10 nm. They have similar physical properties as bulk diamond, such as fluorescence and photoluminescence, as well as biocompatibility. Unlike other CBNs, NDs are made up mostly of tetrahedral clusters of sp-carbon. The surface of nanodiamonds, however, is functionalized with various functional groups or sp-carbon for colloidal stability, which enables chemical modification for targeted drug and gene delivery and tissue labeling.

## Metal-based Nanomaterials

These are metal based materials that we commonly regarded as quantum dots, nanogold, nanosilver and oxides with metal bases. Titanium dioxide is one such example. Metal based Nanomaterials are a focus of the biomedical and pharmaceutical industries. The power here is the chemical binding or conjugated properties that metal-based nanoparticles offer. That power if found in the ability of multi-bond materials to be joined chemically with antibodies or pharmaceuticals.

Preparation of metal-based nanomaterials synthesis through a variety of means such as micro emulsions. The main technique used to create magnetic metal-based nanoparticles is through the manipulation of iron salts via chemical coprecipitation. Metal based nanomaterials are used in healthcare such as contrast dyes that work with MRI and scanning devices for diagnostic purposes.

## Iron Oxide Nanoparticles

Iron (III) oxide ($Fe_2O_3$) is a reddish brown, inorganic compound which is paramagnetic in nature and also one of the three main oxides of iron, while other two being FeO and $Fe_3O_4$. The $Fe_3O_4$, which also occurs naturally as the mineral magnetite, is also super paramagnetic in nature. Due to their ultrafine size, magnetic properties, and biocompatibility, super paramagnetic iron oxide nanoparticles (SPION) have emerged as promising candidates for various biomedical applications, such as enhanced resolution contrast agents for MRI, targeted drug delivery and imaging, hyperthermia, gene therapy, stem cell tracking, molecular/cellular tracking, magnetic separation technologies (e.g., rapid DNA sequencing) early detection of inflammatory, cancer, diabetes, and atherosclerosis. All these biomedical applications require that the nanoparticles have high magnetization values so as to provide high-resolution MR images. In general, the super paramagnetic nanoparticles resemble excellent imaging probes to be used as MRI contrast agents since the MR signal intensity is significantly modulated without any compromise in its in vivo stability. Basically, all contrast agents induce a decrease in the T1 and T2 relaxation times of surrounding water protons and thereby manipulate the signal intensity of the imaged tissue.

Converging advances in the understanding of the molecular biology of various diseases recommended the need of homogeneous and targeted imaging probes along with a narrow size distribution in between 10 and 250 nm in diameter. Developing magnetic nanoparticles in this diameter range is a complex process and various chemical routes for their synthesis have been proposed. These methods include micro emulsions, sol–gel syntheses, sonochemical reactions, hydrothermal reactions, hydrolysis and thermolysis of precursors, flow injection syntheses, and electrospray syntheses. However, the most common method for the production of magnetite nanoparticles is the chemical coprecipitation technique of iron salts. The main advantage of the coprecipitation process is that a large amount of nanoparticles can be synthesized but with limited control on size distribution. This is mainly due to that the kinetic factors are controlling the growth of the crystal. Thus the particulate magnetic contrast agents synthesized using these methods include ultra-small particles of iron oxide (USPIO) (10–40 nm), small particles of iron oxide (SPIO) (60–150 nm). Besides, monocrystalline USPIOs are also called as monocrystalline iron oxide nanoparticles (MIONs), whereas MIONs when cross-linked with dextran they are called crosslinked iron oxide nanoparticles CLIO (10–30 nm). The modification of the dextran coating by carboxylation leads to a shorter clearance half-life in blood. Hence, ferumoxytol (AMAG Pharmaceuticals), a carboxyalkylated polysaccharide coated iron oxide nanoparticle, is already described as a good first-pass contrast agent but uptake by macrophages is unspecific and too fast to enhance the uptake in macrophage-rich plaques.

In order to improve the cellular uptake, these particles can be modified with a peculiar surface coating so that they can be easily conjugated to drugs, proteins, enzymes, antibodies, or nucleotides and can be directed to an organ, tissue, or tumor. While traditional contrast agents distribute rather nonspecifically, targeted molecular imaging probes based on iron oxide nanoparticles have been developed that specifically target body tissue or cells. For instance, Conroy and coworkers developed (chlorotoxin (CTX)) a biocompatible iron oxide nanoprobe coated with poly (ethylene glycol) (PEG), which is capable of specifically targeting glioma tumors via the surface-bound targeting peptide. Further, MRI studies showed the preferential accumulation of the nanoprobe within gliomas. In another study, Apopa engineered iron oxide nanoparticles that can induce an increase in cell permeability through the production of reactive oxygen species (ROS) and the stabilization of microtubules. These are the few applications of iron oxide nanoparticles in biomedical imaging.

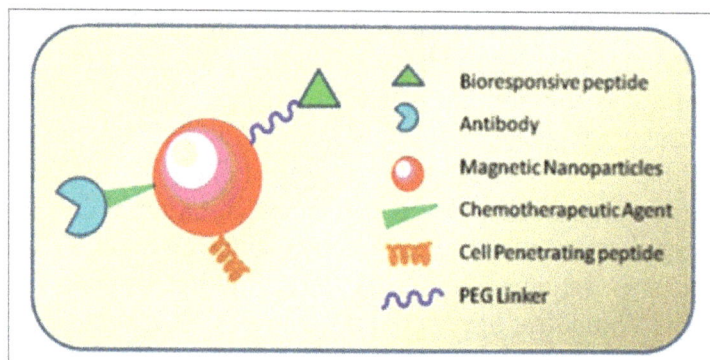

figure: Schematic diagram representing the fucntionalization of magnetic nanoparticles with bioresponsive peptide, PEG linker, chemotherapeutic agent, antibody, and cell-penetrating peptide.

These studies provide a new insight into the bio reactivity of engineered iron nanoparticles, which can provide potential applications in medical imaging or drug delivery. The further development and modification of the complexes of iron oxide along with dendrimers, polymeric nanoparticles, liposomes, and solid lipid nanoparticles are widely studied. However, the toxicity of these magnetic nanoparticles to certain types of neuronal cells is still the matter of concern.

## Gold Nanoparticles

Colloidal gold, also known as gold nanoparticles, is a suspension (or colloid) of nanometer-sized particles of gold. The history of these colloidal solutions dates back to Roman times when they were used to stain glass for decorative purposes. However, the modern scientific evaluation of colloidal gold did not begin until Michael Faraday's work of the 1850s, when he observed that the colloidal gold solutions have properties that differ from the bulk gold. Hence the colloidal solution is either an intense red color (for particles less than 100 nm) or a dirty yellowish color (for larger particles). These interesting optical properties of these gold nanoparticles are due to their unique interaction with light. In the presence of the oscillating electromagnetic field of the light, the free electrons of the metal nanoparticles undergo an oscillation with respect to the metal lattice. This process is resonant at a particular frequency of the light and is termed the localized surface plasmon resonance (LSPR). After absorption, the surface plasmon decays radiatively resulting in light scattering or nonradiatively by converting the absorbed light into heat. Thus for gold nanospheres with particle size around 10 nm in diameter have a strong absorption maximum around 520 nm in aqueous solution due to their LSPR. These nanoshperes show a stokes shift with an increase in the nanosphere size due to the electromagnetic retardation in larger particles.

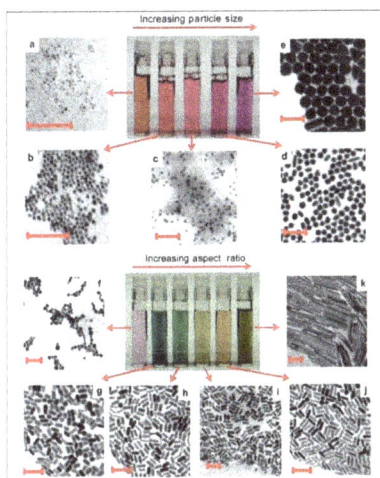

Photographs of aqueous solutions of gold nanospheres (upper panels) and gold nanorods (lower panels) as a function of increasing dimensions. Corresponding transmission electron microscopy images of the nanoparticles are shown (all scale bars 100 nm). The difference in color of the particle solutions is more dramatic for rods than for spheres. This is due to the nature of plasmon bands (one for spheres and two for rods) that are more sensitive to size for rods compared with spheres. For spheres, the size varies from 4 to 40 nm (TEMs a-e), whereas for rods, the aspect ratio varies from 1.3 to 5 for short rods (TEMs f-j) and 20 (TEM k) for long rods.

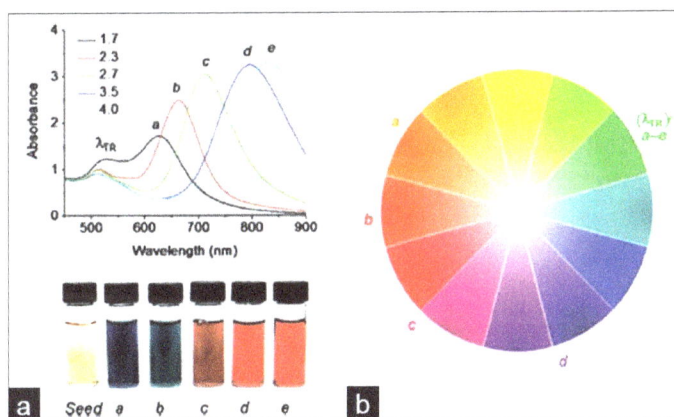

Figure: Gold nanorods (NRs) with tunable optical absorptions at visible and near-infrared wavelengths; a) Optical absorption spectra of gold NRs with different aspect ratios (a–e); b) Color wheel, with reference to gold NRs labeled a–e. TR, transverse resonance

Moreover, the properties and applications of colloidal gold nanoparticles also depend upon its shape. Figure above shows that the difference in color of the particle solutions is more dramatic for rods than for spheres. For example, the rod-shaped nanoparticles have two resonances: one due to plasmon oscillation along the nanorod short axis and another due to plasmon oscillation along the long axis, which depends strongly on the nanorod aspect ratio, that is, length-to-width ratio. When the nanorod aspect ratio is increased, the long-axis LSPR wavelength position red shifts from the visible to the NIR and also progressively increases in oscillator strength. For example, rodlike particles have both transverse and longitudinal absorption peak, and anisotropy of the shape affects their self-assembly. Due to these unique optical properties, gold nanoparticles are the subject of substantial research, with enormous applications including biological imaging, electronics, and materials science. Thus to develop gold nanoparticles for specific applications, reliable and high-yielding methods including those with spherical and nonspherical shapes have been developed over the period of years.

The most prevalent method for the synthesis of monodisperse spherical gold nanoparticles was pioneered by Turkevich in 1951 and later refined by Frens in 1973. This method uses the chemical reduction of gold salts such as hydrogen tetrachloroaurate (HAuCl4) using citrate as the reducing agent. This method produces monodisperse spherical gold nanoparticles in the range of 10–20 nm in diameter. However, the synthesis of larger gold nanoparticles with diameters between 30 and 100 nm was reported by Brown and Natan via seeding of Au3+ by hydroxylamine. Subsequent research led to the modification of the shape of these gold nanoparticles resulting in rod, triangular, polygonal rods, and spherical particles. These ensuing gold nanoparticles have unique properties, providing a high surface area to volume ratio. Moreover, the gold surface offers a unique opportunity to conjugate ligands such as oligonucleotides, proteins, and antibodies containing functional groups such as thiols, mercaptans, phosphines, and amines, which demonstrates a strong affinity for gold surface. The realization of such gold nanoconjugates coupled with strongly enhanced LSPR gold nanoparticles have found applications in simpler but much powerful imaging techniques such as dark-field imaging, SERS, and optical imaging for the diagnosis of various disease states.

In fact, El Sayed have established the use of gold nanoparticles for cancer imaging by selectively transporting AuNPs into the cancer cell nucleus. In order to selectively transport the AuNPs into the cancer cell nucleus, they conjugated arginine–glycine–aspartic acid peptide (RGD) and a nuclear localization signal peptide (NLS) to a 30-nm AuNPs via PEG. RGD is known to target $\alpha v\beta 6$ integrins receptors on the surface of the cell, whereas NLS sequence lysine–lysine–lysine–arginine–lysine (KKKRK) sequence is known to associate with karyopherins (importins) in the cytoplasm, which enables the translocation to the nucleus. Thus the presence of RGD will enable cancer-cell-specific targeting, whereas the presence of NLS will exhibit cancer cell nucleus specific targeting. This intuitively developed particle was then targeted to human oral squamous cell carcinoma (HSC) having $\alpha v\beta 6$ integrins overexpressed on the cell surface (cancer model), and human keratinocytes (HaCat) (control). The authors further demonstrated that RGD-AuNPs specifically target the cytoplasm of cancer cells over that of normal cells figure c, and the RGD/NLS-AuNPs specifically target the nuclei of cancer cells figure b, over those of normal cells.

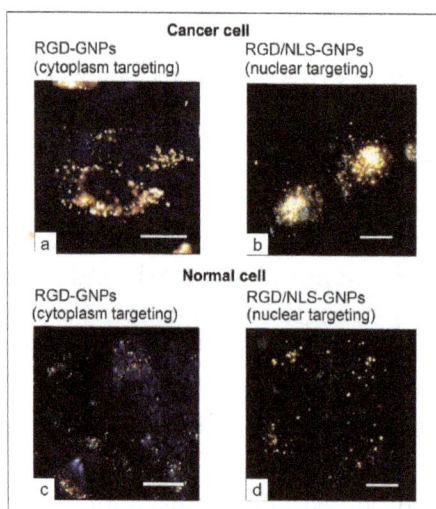

Dark field light scattering images of cytoplasm and nuclear targeting AuNPs. a) RGD-AuNPs located in the cytoplasm of cancer cells. b) RGD/NLS-AuNPs located at the nucleus of cancer cells. c) RGD-AuNPs located in the cytoplasm of normal cells. d) RGD/NLS-AuNPs located at the nucleus of normal cells. The cancer and normal cells were incubated in the presence of these AuNPs at a concentration of 0.4 nM for 24 hours and these images clearly display the efficient uptake of AuNPs in cancer cells compared with normal cells. Scale bar 10 μm.

Similarly, Qian reported the development of tumor-targeted gold nanoparticles as a probe for Raman scatters in vivo. These gold nanoparticles were encoded with a Raman reporter and further encapsulated with a thiol-modified PEG coat. Additionally, to specifically target tumor cells, the pegylated gold nanoparticles were then conjugated with an antibody against epidermal growth factor receptor, which is sometimes overexpressed in certain types of cancer cells. The Raman enhancement from these tailored particles was then observed with electronic transitions at 633 or 785 nm via SERS. The results obtained by Qian and coworkers suggest the highly specific recognition and detection of human cancer cells, as well as active targeting of EGFR-positive tumor xenografts in animal models can be made using SERS.

Moreover, the use of gold nanorods as photothermal agents sets them apart from all nanoprobes. Photothermal therapy (PTT) is a procedure in which a photosensitizer is excited with specific band light (mainly IR). This activation brings the sensitizer to an excited state where it then releases vibrational energy in the form of heat. The heat is the actual method of therapy that kills the targeted cells. One of the biggest recent successes in photothermal therapy is the use of gold nanoparticles. Spherical gold nanoparticles absorptions have not been optimal for in vivo applications. This is because the peak absorptions have been limited to 520 nm for 10 nm diameter. Moreover, skin, tissues, and hemoglobin have a transmission window from 650 up to 900 nm. This was circumvented by the recent invention of gold nanorods by Murphy and Coworkers, who were able to tune the absorption peak of these nanoparticles, which can also be tuned from 550 nm up to 1 μm just by altering its aspect ratio of the nanorods. Hence, for the rod-shaped gold nanoparticles with the absorption in the IR region, when selectively accumulated in tumors when bathed in laser light (in the IR region), the surrounding tissue is barely warmed, but the nanorods convert light to heat, killing the malignant cells. This potential application of gold nanorods sanctifies them from other nanoprobes. However, their incompatibility with other high-resolution imaging techniques such as MRI and irreproducibility in shapes led to the invention of nanocages and nanoshells.

## Nanoshells and Nanocages

Neeves and Birnboim calculated that a composite spherical particle consisting of a metallic shell and a dielectric core could give rise to LSPR modes with their wavelengths tunable over a broad range of the electromagnetic spectrum. Later on, the experimental and theoretical work by Naomi Halas and Peter Nordlander at Rice University showed that the resonance of a silica-gold nanoshell particle can easily be positioned in the near-infrared (800–1,300 nm) region, where absorption by biomatters is low. They developed silica-gold nanospheres by using freshly formed amine-terminated silica spheres. These amine terminated silica spheres were then treated with a suspension of gold colloids (1–2 nm in size). Gold was deposited via chemical reduction to cover the silica core and to the amine terminal of the silicon core. Although this method is widely used, the intricacy involved in the control of thickness and smoothness of the metallic shells makes this method unsuitable for the routine synthesis of controlled particle-sized nanoshells. Furthermore, they also showed the successful irreversible photothermal ablation of tumor tissue both in vitro and in vivo when these nanoshells were localized onto the tumor cells. In another study, Halas and West established the use of near-infrared resonant nanoshells for whole-blood immunoassays. They further showed that the nanoshells when conjugated with antibodies act as recognition sites for a specific analyte. The analyte causes the formation of dimmers, which will modify the LSPR. Subsequent work in this field led to the development of the multifunctional magnetic gold nanoshells (Mag-GNS) by Jaeyun utilizing $Fe_3O_4$ nanoparticles as the magnetic core. The $Fe_3O_4$ nanoparticles allow MRI for diagnosis, and the gold nanoshells enable photothermal therapy. By attaching an antibody to the Mag-GNS by a PEG linker, cancer cells can be targeted. Once localized, these particles enable the detection of cancer using MRI, whereas the photothermal therapy can be used to get rid of cancer cells.

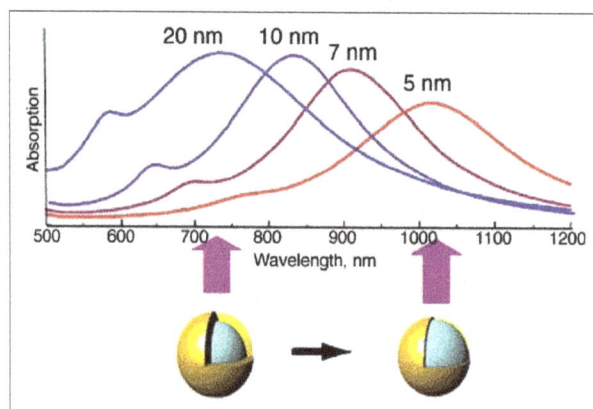

Figure: Gold nanoshell plasmon resonances for a 120-nm core with indicated shell thickness

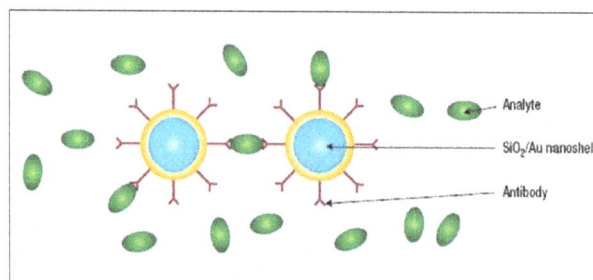

Similar to gold nanoshells, gold nanocages represent a novel class of nanostructures that are hollow porous gold nanoparticles that absorb light in the near-infrared range. They were first developed by the Xia and Coworkers via the reaction of silver nanoparticles with chloroauric acid (HAu-CI4) in boiling water. Their LSPR peaks can also be tuned to the near infrared region by controlling the thickness and porosity of the walls. Comparable to nanoshells they have found applications in drug delivery and/or controlled drug release. Furthermore, the hollow interiors can host small objects such as magnetic nanoparticles to construct multifunctional hybrid nanostructures diagnostic imaging and therapy.

## Silver Nanoparticles

Silver nanoparticles are the particles of silver, with particle size between 1 and 100 nm in size. While frequently described as being "silver" some are composed of a large percentage of silver oxide due to their large ratio of surface to bulk silver atoms. Like gold nanoparticles, ionic silver has a long history and was initially used to stain the glass for yellow. Currently, there is also an effort to incorporate silver nanoparticles into a wide range of medical devices, including bone cement, surgical instruments, surgical masks, etc. Moreover, it has also been shown that ionic silver, in the right quantities, is suitable in treating wounds. In fact, silver nanoparticles are now replacing silver sulfadiazine as an effective agent in the treatment of wounds. Additionally, Samsung has created and marketed a material called Silver Nano, which includes silver nanoparticles on the surfaces of household appliances. Moreover, due to their attractive physiochemical properties these nanomaterials have received considerable attention in biomedical imaging using SERS. In fact, the surface plasmon resonance and large effective scattering cross-section of individual silver nanoparticles make them ideal candidates for molecular labeling. Thus many targeted silver oxide nanoprobes are currently being developed.

Typically, they are synthesized by the reduction of a silver salt with a reducing agent like sodium borohydride in the presence of a colloidal stabilizer. The most common colloidal stabilizers used are polyvinyl alcohol, poly (vinylpyrrolidone), bovine serum albumin (BSA), citrate, and cellulose. Newer novel methods include the use of β-d-glucose as a reducing sugar and a starch as the stabilizer to develop silver nanoparticles ion implantation used to create silver nanoparticles. Also, it is important to note that not all nanoparticles created are equal. The size and shape have been shown to have an impact on its efficacy. In fact, Elechiguerra et al. demonstrated that silver nanoparticles undergo a size-dependent interaction with HIV-1, with particles exclusively in the range of 1–10 nm attached to the virus. They further suggest that silver nanoparticles interact with the HIV-1 virus via preferential binding to the gp120 glycoprotein knobs. Similarly, Furno and Coworkers have developed biomaterials by impregnating silicone coated with silver oxide nanoparticles using supercritical carbon dioxide. These novel biomaterials were developed with an aim to reduce the antibacterial infection. The results obtained were mixed but the methodology allows for the first-time silver impregnation (as opposed to coating) of medical polymers and promises to lead to an antimicrobial biomaterial.

Even though these particles are not as widely preferred as compared to the gold nanoparticles and nanoshells, but they have made a tremendous impact on today's era of medical science. The interesting property of the noble metals is a promise that they would be continuously used as newer applications and protocols are being developed.

## Dendrimers

Dendrimers are nano-sized, radially symmetric molecules with well-defined, homogeneous, and monodisperse structure consisting of tree-like arms or branches) line then tk from (The second group called synthesized macromolecules 'arborols' means, in Latin, 'trees.

The second group called synthesized macromolecules 'arborols' means, in Latin, 'trees'. Dendrimers might also be called 'cascade molecules', but this term is not as much established as 'dendrimers'. Dendrimers are nearly monodisperse macromolecules that contain symmetric branching units built around a small molecule or a linear polymer core. 'Dendrimer' is only an architectural motif and not a compound. Polyionic dendrimers do not have a persistent shape and may undergo changes in size, shape, and flexibility as a function of increasing generations. Dendrimers are hyperbranched macromolecules with a carefully tailored architecture, the end-groups (i.e., the groups reaching the outer periphery), which can be functionalized, thus modifying their physicochemical or biological properties. Dendrimers have gained a broad range of applications in supramolecular chemistry, particularly in host-guest reactions and self-assembly processes. Dendrimers are characterized by special features that make them promising candidates for a lot of applications. Dendrimers are highly defined artificial macromolecules, which are characterized by a combination of a high number of functional groups and a compact molecular structure. The emerging role of dendritic macromolecules for anticancer therapies and diagnostic imaging is remarkable. The advantages of these well-defined materials make them the newest class of macromolecular nano-scale delivery devices. Dendritic macromolecules tend to linearly increase in diameter and adopt a more globular shape with increasing dendrimer generation. Therefore, dendrimers have become an ideal delivery vehicle candidate for explicit study of the effects of polymer size, charge, and composition on biologically relevant properties such as lipid bilayer interactions, cytotoxicity, internalization, blood plasma retention time, biodistribution, and filtration.

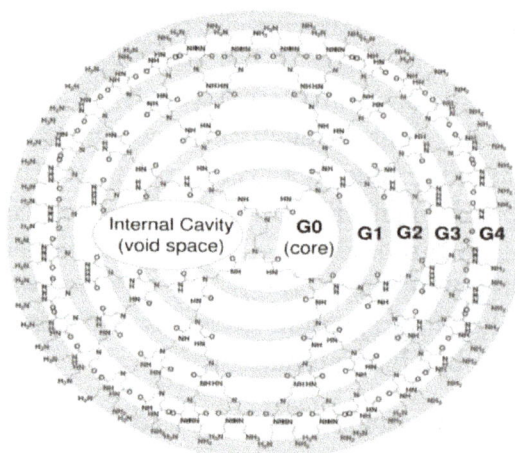

Schematic representation of a generation G4 dendrimer with 64 amino groups at the periphery. This dendrimer starts from an ethylene diamine core; the branches or arms were attached by exhaustive Michael addition to methyl acrylate followed by exhaustive aminolysis of the resulting methyl ester using ethylene diamine.

## Structure and Chemistry

The structure of dendrimer molecules begins with a central atom or group of atoms labeled as the core. From this central structure, the branches of other atoms called 'dendrons' grow through a variety of chemical reactions. There continues to be a debate about the exact structure of dendrimers, in particular whether they are fully extended with maximum density at the surface or whether the end-groups fold back into a densely packed interior. Dendrimers can be prepared with a level of control not attainable with most linear polymers, leading to nearly monodisperse, globular macromolecules with a large number of peripheral groups as seen in figure below, the structure of some dendrimer repeat units, for example, the 1,3-diphenylacetylene unit developed by Moore.

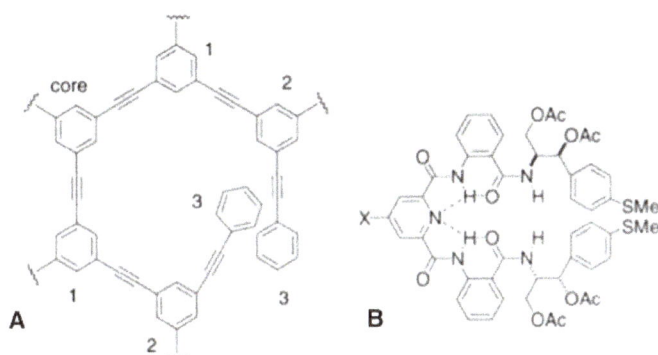

Types of dendrimers. (A) More type dendrimers consisting of phenyl acetylene subunits at the third-generation different arms may dwell in the same space, and the fourth-generation layer potential overlaps with the second-generation layer. (B) Parquette-type dendrons are chiral, non-racemic, and with intramolecular folding driven by hydrogen bonding.

Dendrimers are a new class of polymeric belongings. Their chemistry is one of the most attractive and hastily growing areas of new chemistry. Dendrimer chemistry, as other specialized research fields, has its own terms and abbreviations. Dendrigrafts are a class of dendritic polymers like dendrimers that can be constructed with a well-defined molecular structure, i.e., being monodisperse. The unique structure of dendrimers provides special opportunities for host-guest chemistry and is especially well equipped to engage in multivalent interactions. At the same time, one of the first proposed applications of dendrimers was as container compounds, wherein small substrates are bound within the internal voids of the dendrimer. Experimental evidence for unimolecular micelle properties was established many years ago both in hyperbranched polymers and dendrimers.

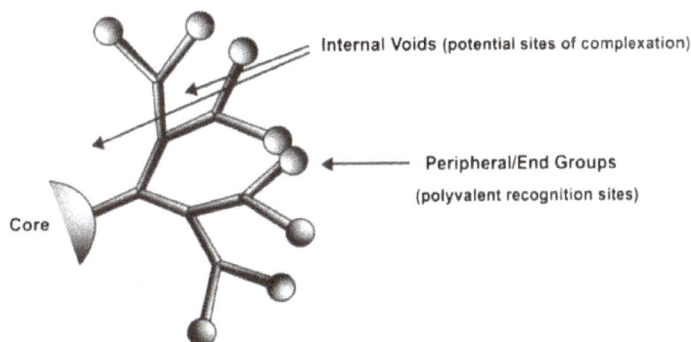

## Synthesis

One of the very first dendrimers, the Newkome dendrimer, was synthesized in 1985. This macromolecule is also commonly known by the name arborol. The figure outlines the mechanism of the first two generations of arborol through a divergent route. The synthesis is started by nucleophilic substitution of 1-bromopentane by triethyl sodiomethanetricarboxylate in dimethylformamide and benzene. The ester groups were then reduced by lithium aluminium hydride to a triol in a deprotection step. Activation of the chain ends was achieved by converting the alcohol groups to tosylate groups with tosyl chloride and pyridine. The tosyl group then served as leaving groups in another reaction with the tricarboxylate, forming generation two. Further repetition of the two steps leads to higher generations of arborol.

Poly (amidoamine), or PAMAM, is perhaps the most well known dendrimer. The core of PAMAM is a diamine (commonly ethylenediamine), which is reacted with methyl acrylate, and then another ethylenediamine to make the generation-0 (G-0) PAMAM. Successive reactions create higher generations, which tend to have different properties. Lower generations can be thought of as flexible molecules with no appreciable inner regions, while medium-sized (G-3 or G-4) do have internal space that is essentially separated from the outer shell of the dendrimer. Very large (G-7 and greater) dendrimers can be thought of more like solid particles with very dense surfaces due to the structure of their outer shell. The functional group on the surface of PAMAM dendrimers is ideal for click chemistry, which gives rise to many potential applications.

Dendrimers can be considered to have three major portions: a core, an inner shell, and an outer shell. Ideally, a dendrimer can be synthesized to have different functionality in each of these portions to control properties such as solubility, thermal stability, and attachment of compounds for particular applications. Synthetic processes can also precisely control the size and number of branches on the dendrimer. There are two defined methods of dendrimer synthesis, divergent synthesis and convergent synthesis. However, because the actual reactions consist of many steps needed to protect the active site, it is difficult to synthesize dendrimers using either method. This makes dendrimers hard to make and very expensive to purchase. At this time, there are only a few companies that sell dendrimers; Polymer Factory Sweden AB commercializes biocompatible bis-MPA dendrimers and Dendritech is the only kilogram-scale producers of PAMAM dendrimers. NanoSynthons, LLC from Mount Pleasant, Michigan, USA produces PAMAM dendrimers and other proprietary dendrimers.

## Divergent Methods

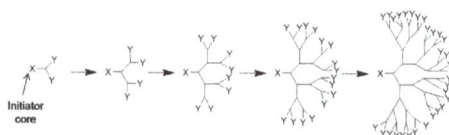

Schematic of divergent synthesis of dendrimers

The dendrimer is assembled from a multifunctional core, which is extended outward by a series of reactions, commonly a Michael reaction. Each step of the reaction must be driven to full completion to prevent mistakes in the dendrimer, which can cause trailing generations (some branches are shorter than the others). Such impurities can impact the functionality and symmetry of the dendrimer, but are extremely difficult to purify out because the relative size difference between perfect and imperfect dendrimers is very small.

## Convergent Methods

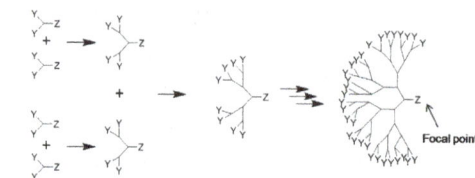

Schematic of convergent synthesis of dendrimers

Dendrimers are built from small molecules that end up at the surface of the sphere, and reactions proceed inward building inward and are eventually attached to a core. This method makes it much easier to remove impurities and shorter branches along the way, so that the final dendrimer is more monodisperse. However dendrimers made this way are not as large as those made by divergent methods because crowding due to steric effects along the core is limiting.

## Click Chemistry

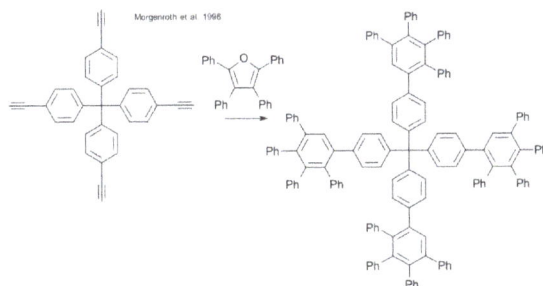

Dendrimer Diels-Alder reaction.

Dendrimers have been prepared via click chemistry, employing Diels-Alder reactions, thiol-ene and thiol-yne reactions and azide-alkyne reactions.

There are ample avenues that can be opened by exploring this chemistry in dendrimer synthesis.

## Applications

Applications of dendrimers typically involve conjugating other chemical species to the dendrimer surface that can function as detecting agents (such as a dye molecule), affinity ligands, targeting components, radioligands, imaging agents, or pharmaceutically active compounds. Dendrimers have very strong potential for these applications because their structure can lead to multivalent systems. In other words, one dendrimer molecule has hundreds of possible sites to couple to an active species. Researchers aimed to utilize the hydrophobic environments of the dendritic media to conduct photochemical reactions that generate the products that are synthetically challenged. Carboxylic acid and phenol-terminated water-soluble dendrimers were synthesized to establish their utility in drug delivery as well as conducting chemical reactions in their interiors. This might allow researchers to attach both targeting molecules and drug molecules to the same dendrimer, which could reduce negative side effects of medications on healthy cells.

Dendrimers can also be used as a solubilizing agent. Since their introduction in the mid-1980s, this novel class of dendrimer architecture has been a prime candidate for host-guest chemistry. Dendrimers with hydrophobic core and hydrophilic periphery have shown to exhibit micelle-like behavior and have container properties in solution. The use of dendrimers as unimolecular micelles was proposed by Newkome in 1985. This analogy highlighted the utility of dendrimers as solubilizing agents. The majority of drugs available in pharmaceutical industry are hydrophobic in nature and this property in particular creates major formulation problems. This drawback of drugs can be ameliorated by dendrimeric scaffolding, which can be used to encapsulate as well as to solubilize the drugs because of the capability of such scaffolds to participate in extensive hydrogen bonding with water. Dendrimer labs throughout the planet are persistently trying to manipulate dendrimer's solubilizing trait, in their way to explore dendrimer as drug delivery and target specific carrier.

For dendrimers to be able to be used in pharmaceutical applications, they must surmount the required regulatory hurdles to reach market. One dendrimer scaffold designed to achieve this is the Poly Ethoxy Ethyl Glycinamide (PEE-G) dendrimer. This dendrimer scaffold has been designed and shown to have high HPLC purity, stability, aqueous solubility and low inherent toxicity.

## Drug Delivery

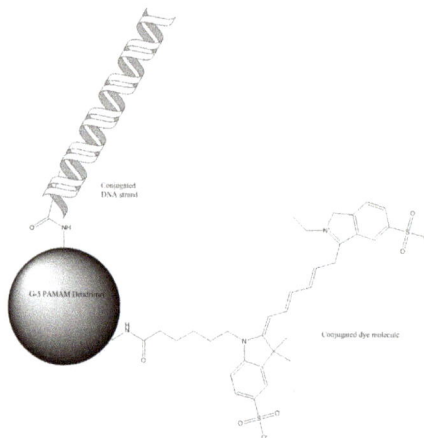

Schematic of a G-5 PAMAM dendrimer conjugated to both a dye molecule and a strand of DNA.

Approaches for delivering unaltered natural products using polymeric carriers is of widespread interest, dendrimers have been explored for the encapsulation of hydrophobic compounds and for the delivery of anticancer drugs. The physical characteristics of dendrimers, including their monodispersity, water solubility, encapsulation ability, and large number of functionalizable peripheral groups, make these macromolecules appropriate candidates for evaluation as drug delivery vehicles. There are three methods for using dendrimers in drug delivery: first, the drug is covalently attached to the periphery of the dendrimer to form dendrimer prodrugs, second the drug is coordinated to the outer functional groups via ionic interactions, or third the dendrimer acts as a unimolecular micelle by encapsulating a pharmaceutical through the formation of a dendrimer-drug supramolecular assembly. The use of dendrimers as drug carriers by encapsulating hydrophobic drugs is a potential method for delivering highly active pharmaceutical compounds that may not be in clinical use due to their limited water solubility and resulting suboptimal pharmacokinetics. Dendrimers have been widely explored for controlled delivery of antiretroviral bioactives The inherent antiretroviral activity of dendrimers enhances their efficacy as carriers for antiretroviral drugs. The dendrimer enhances both the uptake and retention of compounds within cancer cells, a finding that was not anticipated at the onset of studies. The encapsulation increases with dendrimer generation and this method may be useful to entrap drugs with a relatively high therapeutic dose. Studies based on this dendritic polymer also open up new avenues of research into the further development of drug-dendrimer complexes specific for a cancer and/or targeted organ system. These encouraging results provide further impetus to design, synthesize, and evaluate dendritic polymers for use in basic drug delivery studies and eventually in the clinic.

## Gene Delivery

The ability to deliver pieces of DNA to the required parts of a cell includes many challenges. Current research is being performed to find ways to use dendrimers to traffic genes into cells without damaging or deactivating the DNA. To maintain the activity of DNA during dehydration, the dendrimer/DNA complexes were encapsulated in a water-soluble polymer, and then deposited on or sandwiched in functional polymer films with a fast degradation rate to mediate gene transfection. Based on this method, PAMAM dendrimer/DNA complexes were used to encapsulate functional biodegradable polymer films for substratemediated gene delivery. Research has shown that the fast-degrading functional polymer has great potential for localized transfection.

## Sensors

Dendrimers have potential applications in sensors. Studied systems include proton or pH sensors using poly(propylene imine), cadmium-sulfide/polypropylenimine tetrahexacontaamine dendrimer composites to detect fluorescence signal quenching, and poly(propylenamine) first and second generation dendrimers for metal cation photodetection amongst others. Research in this field is vast and ongoing due to the potential for multiple detection and binding sites in dendritic structures.

## Blood Substitution

Dendrimers are also being investigated for use as blood substitutes. Their steric bulk surrounding a heme-mimetic centre significantly slows degradation compared to free heme, and prevents the cytotoxicity exhibited by free heme.

## Nanoparticles

Dendrimers also are used in the synthesis of monodisperse metallic nanoparticles. Poly(amidoamide), or PAMAM, dendrimers are utilized for their tertiary amine groups at the branching points within the dendrimer. Metal ions are introduced to an aqueous dendrimer solution and the metal ions form a complex with the lone pair of electrons present at the tertiary amines. After complexion, the ions are reduced to their zerovalent states to form a nanoparticle that is encapsulated within the dendrimer. These nanoparticles range in width from 1.5 to 10 nanometers and are called dendrimer-encapsulated nanoparticles.

## Crop Protection and Agrochemicals

Given the widespread use of pesticides, herbicides and insecticides in modern farming, dendrimers are also being used by companies to help improve the delivery of agrochemicals to enable healthier plant growth and to help fight plant diseases.

# Composites

## Composite Material

Composite materials (or composites for short) are engineered materials made from two or more constituent materials with significantly different physical or chemical properties and which remain separate and distinct on a macroscopic level within the finished structure.

A cloth of woven carbon fiber filaments,
a common element in composite materials

Plywood is a common composite material
many people encounter in their everyday life.

The most primitive composite materials comprised straw and mud in the form of bricks for building construction; the Biblical book of Exodus speaks of the Israelites being oppressed by Pharaoh, by being forced to make bricks without straw. The ancient brick-making process can still be seen on Egyptian tomb paintings in the Metropolitan Museum of Art. The most advanced examples perform routinely on spacecraft in demanding environments. The most visible applications pave our roadways in the form of either steel and aggregate reinforced portland cement or asphalt concrete. Those composites closest to our personal hygiene form our shower stalls and bath tubs made of fiberglass. Solid surface, imitation granite and cultured marble sinks and counter tops are widely used to enhance our living experiences.

There are two categories of constituent materials: matrix and reinforcement. At least one portion of each type is required. The matrix material surrounds and supports the reinforcement materials by maintaining their relative positions. The reinforcements impart their special mechanical and physical properties to enhance the matrix properties. A synergism produces material properties unavailable from the individual constituent materials, while the wide variety of matrix and strengthening materials allows the designer of the product or structure to choose an optimum combination.

Engineered composite materials must be formed to shape. The matrix material can be introduced to the reinforcement before or after the reinforcement material is placed into the mold cavity or onto the mold surface. The matrix material experiences a melding event, after which the part shape is essentially set. Depending upon the nature of the matrix material, this melding event can occur in various ways such as chemical polymerization or solidification from the melted state.

A variety of molding methods can be used according to the end-item design requirements. The principal factors impacting the methodology are the natures of the chosen matrix and reinforcement materials. Another important factor is the gross quantity of material to be produced. Large quantities can be used to justify high capital expenditures for rapid and automated manufacturing technology. Small production quantities are accommodated with lower capital expenditures but higher labor and tooling costs at a correspondingly slower rate.

Most commercially produced composites use a polymer matrix material often called a resin solution. There are many different polymers available depending upon the starting raw ingredients. There are several broad categories, each with numerous variations. The most common are known as polyester, vinyl ester, epoxy, phenolic, polyimide, polyamide, polypropylene, PEEK, and others. The reinforcement materials are often fibers but also commonly ground minerals.

## Molding Methods

In general, the reinforcing and matrix materials are combined, compacted and processed to undergo a melding event. After the melding event, the part shape is essentially set, although it can deform under certain process conditions. For a thermoset polymeric matrix material, the melding event is a curing reaction that is initiated by the application of additional heat or chemical reactivity such as an organic peroxide. For a thermoplastic polymeric matrix material, the melding event is a solidification from the melted state. For a metal matrix material such as titanium foil, the melding event is a fusing at high pressure and a temperature near the melt point.

For many molding methods, it is convenient to refer to one mold piece as a "lower" mold and another mold piece as an "upper" mold. Lower and upper refer to the different faces of the molded panel, not the mold's configuration in space. In this convention, there is always a lower mold, and sometimes an upper mold. Part construction begins by applying materials to the lower mold. Lower mold and upper mold are more generalized descriptors than more common and specific terms such as male side, female side, a-side, b-side, tool side, bowl, hat, mandrel, etc. Continuous manufacturing processes use a different nomenclature.

The molded product is often referred to as a panel. For certain geometries and material combinations, it can be referred to as a casting. For certain continuous processes, it can be referred to as a profile.

## Open Molding

A process using a rigid, one sided mold which shapes only one surface of the panel. The opposite surface is determined by the amount of material placed upon the lower mold. Reinforcement materials can be placed manually or robotically. They include continuous fiber forms fashioned into textile constructions and chopped fiber. The matrix is generally a resin, and can be applied with a pressure roller, a spray device or manually. This process is generally done at ambient temperature and atmospheric pressure. Two variations of open molding are Hand Layup and Spray-up.

## Vacuum Bag Molding

A process using a two-sided mold set that shapes both surfaces of the panel. On the lower side is a rigid mold and on the upper side is a flexible membrane. The flexible membrane can be a reusable silicone material or an extruded polymer film such as nylon. Reinforcement materials can be placed on the lower mold manually or robotically, generally as continuous fiber forms fashioned into textile constructions. The matrix is generally a resin. The fiber form may be pre-impregnated with the resin in the form of prepreg fabrics or unidirectional tapes. Otherwise, liquid matrix material is introduced to dry fiber forms prior to applying the flexible film. Then, vacuum is applied to the mold cavity. This process can be performed at either ambient or elevated temperature with ambient atmospheric pressure acting upon the vacuum bag. Most economical way is using a venturi vacuum and air compressor or a vacuum pump.

## Autoclave Molding

A process using a two-sided mold set that forms both surfaces of the panel. On the lower side is a rigid mold and on the upper side is a flexible membrane made from silicone or an extruded polymer film such as nylon. Reinforcement materials can be placed manually or robotically. They include continuous fiber forms fashioned into textile constructions. Most often, they are pre-impregnated with the resin in the form of prepreg fabrics or unidirectional tapes. In some instances, a resin film is placed upon the lower mold and dry reinforcement is placed above. The upper mold is installed and vacuum is applied to the mold cavity. Then, the assembly is placed into an autoclave pressure vessel. This process is generally performed at both elevated pressure and elevated temperature. The use of elevated pressure facilitates a high fiber volume fraction and low void content for maximum structural efficiency.

## Resin Transfer Molding

A process using a two-sided mold set that forms both surfaces of the panel. The lower side is a rigid mold. The upper side can be a rigid or flexible mold. Flexible molds can be made from composite materials, silicone or extruded polymer films such as nylon. The two sides fit together to produce a mold cavity. The distinguishing feature of resin transfer molding is that the reinforcement materials are placed into this cavity and the mold set is closed prior to the introduction of matrix material. Resin transfer molding includes numerous varieties which differ in the mechanics of how the resin is introduced to the reinforcement in the mold cavity. These variations include everything from vacuum infusion to vacuum assisted resin transfer molding. This process can be performed at either ambient or elevated temperature.

## Other

Other types of molding include press molding, transfer molding, pultrusion molding, filament winding, casting, centrifugal casting and continuous casting.

## Tooling

Some types of tooling materials used in the manufacturing of composites structures include invar, steel, aluminum, reinforced silicon rubber, nickle, and carbon fiber. Selection of the tooling material is typically based on, but not limited to, the coefficient of thermal expansion, expected number of cycles, end item tolerance, desired or required surface condition, method of cure, glass transition temperature of the material being molded, molding method, matrix, cost and a variety of other considerations.

## Mechanics of Composite Materials

The physical properties of composite materials are generally not isotropic in nature, but rather are typically orthotropic. For instance, the stiffness of a composite panel will often depend upon the directional orientation of the applied forces and/or moments. Panel stiffness is also dependent on the design of the panel. For instance, the fiber reinforcement and matrix used, the method of panel build, thermoset versus thermoplastic, type of weave, and orientation of fiber axis to the primary force.

In contrast, isotropic materials (for example, aluminum or steel), in standard wrought forms, typically have the same stiffness regardless of the directional orientation of the applied forces and/or moments.

The relationship between forces/moments and strains/curvatures for an isotropic material can be described with the following material properties: Young's Modulus, the Shear Modulus and the Poisson's ratio, in relatively simple mathematical relationships. For the anisotropic material, it requires the mathematics of a second order tensor and can require up to 21 material property constants. For the special case of orthogonal isotropy, there are three different material property constants for each of Young's Modulus, Shear Modulus and Poisson's Ratio for a total of nine material property constants to describe the relationship between forces/moments and strains/curvatures.

## Categories of Fiber Reinforced Composite Materials

Fiber reinforced composite materials can be divided into two main categories normally referred to as short fiber reinforced materials and continuous fiber reinforced materials. Continuous reinforced materials will often constitute a layered or laminated structure. The woven and continuous fiber styles are typically available in a variety of forms, being pre-impregnated with the given matrix (resin), dry, uni-directional tapes of various widths, plain weave, harness satins, braided, and stitched.

The short and long fibers are typically employed in compression molding and sheet molding operations. These come in the form of flakes, chips, and random mate (which can also be made from a continuous fiber laid in random fashion until the desired thickness of the ply/laminate is achieved).

## Failure of Composites

Shock, impact, or repeated cyclic stresses can cause the laminate to separate at the interface between two layers, a condition known as delamination. Individual fibers can separate from the matrix e.g. fiber pull-out.

Composites can fail on the microscopic or macroscopic scale. Compression failures can occur at both the macro scale or at each individual reinforcing fiber in compression buckling. Tension failures can be net section failures of the part or degradation of the composite at a microscopic scale where one or more of the layers in the composite fail in tension of the matrix or failure the bond between the matrix and fibers.

Some composites are brittle and have little reserve strength beyond the initial onset of failure while others may have large deformations and have reserve energy absorbing capacity past the onset of damage. The variations in fibers and matrices that are available and the mixtures that can be made with blends leave a very broad range of properties that can be designed into a composite structure.

## Examples of Composite Materials

Fiber Reinforced Polymers or FRPs include Wood comprising (cellulose fibers in a lignin and hemicellulose matrix), Carbon-fiber reinforced plastic or CFRP, Glass-fiber reinforced plastic or GFRP (also GRP). If classified by matrix then there are Thermoplastic Composites, short fiber thermoplastics, long fiber thermoplastics or long fiber reinforced thermoplastics There are numerous thermoset composites, but advanced systems usually incorporate aramid fiber and carbon fiber in an epoxy resin matrix.

Composites can also utilize metal fibers reinforcing other metals, as in Metal matrix composites or MMC. Ceramic matrix composites include Bone (hydroxyapatite reinforced with collagen fibers), Cermet (ceramic and metal) and Concrete. Organic matrix/ceramic aggregate composites include Asphalt concrete, Mastic asphalt, Mastic roller hybrid, Dental composite, Syntactic foam and Mother of Pearl. Chobham armour is a special composite used in military applications.

Additionally, thermoplastic composite materials can be formulated with specific metal powders resulting in materials with a density range from two g/cc to 11 g/cc (same density as lead). These materials can be used in place of traditional materials such as aluminum, stainless steel, brass, bronze, copper, lead, and even tungsten in weighting, balancing, vibration dampening, and radiation shielding applications. High density composites are an economically viable option when certain materials are deemed hazardous and are banned (such as lead) or when secondary operations costs (such as machining, finishing, or coating) are a factor.

Engineered wood includes a wide variety of different products such as Plywood, Oriented strand board, Wood plastic composite (recycled wood fiber in polyethylene matrix), Pykrete (sawdust in ice matrix), Plastic-impregnated or laminated paper or textiles, Arborite, Formica (plastic) and Micarta. Other engineered laminate composites, such as Mallite, use a central core of end grain balsa wood, bonded to surface skins of light alloy or GRP. These generate low-weight, high rigidity materials.

## Typical Products

Composite materials have gained popularity (despite their generally high cost) in high-performance products such as aerospace components (tails, wings , fuselages, propellors), boat and scull hulls, and racing car bodies. More mundane uses include fishing rods and storage tanks.

## References

- Vögtle, Fritz / Richardt, Gabriele / Werner, Nicole Dendrimer Chemistry Concepts, Syntheses, Properties, Applications 2009 ISBN 3-527-32066-0

- Buhleier, Egon; Wehner, Winfried; Vögtle, Fritz (1978). ""Cascade"- and "Nonskid-Chain-like" Syntheses of Molecular Cavity Topologies". Synthesis. 1978 (2): 155–158. doi:10.1055/s-1978-24702

- Composite-material: newworldencyclopedia.org, Retrieved 28 May 2018

- Holister, Paul; Christina Roman Vas; Tim Harper (October 2003). "Dendrimers: Technology White Papers" (PDF). Cientifica. Archived from the original (PDF) on 6 July 2011. Retrieved 17 March 2010

- Franc, Grégory; Kakkar, Ashok K. (2009). "Diels-Alder "Click" Chemistry in Designing Dendritic Macromolecules". Chemistry: A European Journal. 15 (23): 5630–5639. doi:10.1002/chem.200900252.

# 6

# Nanometrology

Nanometrology is a subfield of metrology, which involves the science of measurement at nanoscales. This chapter discusses the diverse measurement techniques such as atomic force microscopy, scanning tunneling microscopy, etc. for a holistic understanding of nanometrology.

Nanometrology involves the measurement of geometrical features of size, shape and roughness at the nanoscale. Though not an innate aspect of the specimen under study, these geometrical features are often measured against an arbitrarily fixed co-ordinate system, more so in an engineering application. With advent of new age metrological instruments (e.g. SPM), the parameters of measurement also include a physical quantity (e.g. force). Another important aspect that has to be borne in mind when the scale of objects reduces is the relative importance or relevance of the physical features namely size, shape and roughness. Below are three plots that illustrate the relative importance of the physical features as the scale reduces.

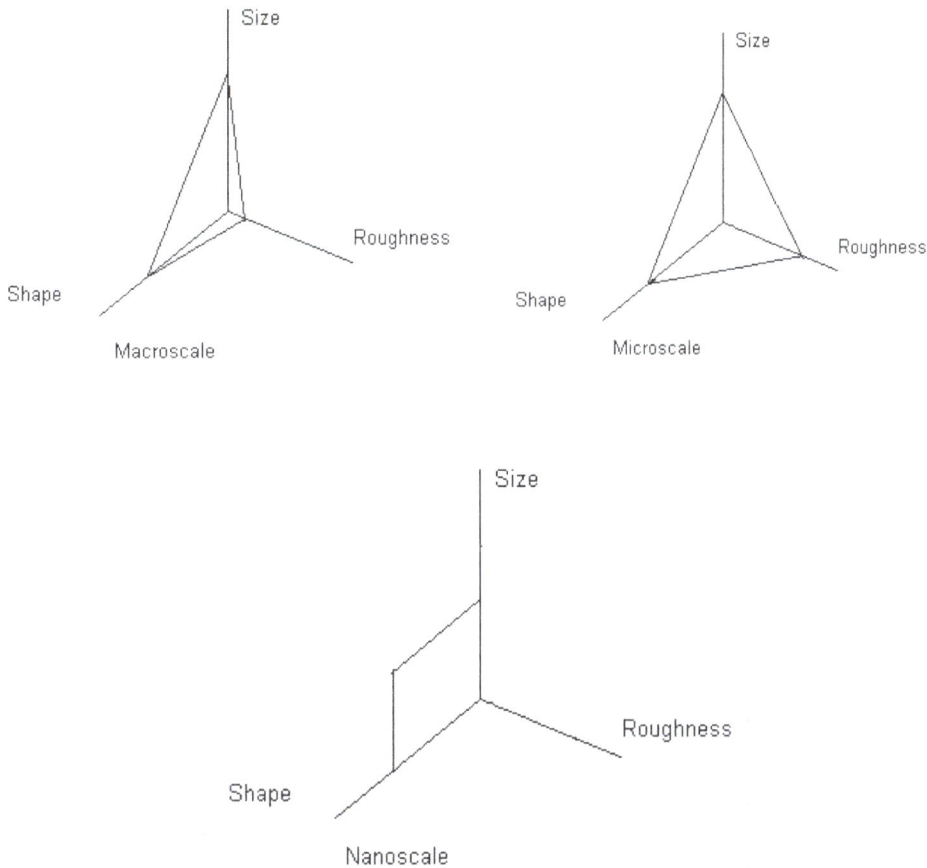

Figure: Relative relationship between different geometric components in conventional, micromechanics & quantum engineering.

Macro scale measurements are usually encountered in conventional engineering applications; micro scale measurements are encountered in micro mechanics while nanoscale measurements are encountered in quantum engineering. Also, as scale reduces additional physical constraints such as tunneling and other such related phenomenon are also to be considered. The above plots, in a crude way, explain how the very definition of roughness changes as we reduce the scale.

## At the Nanoscale

In conventional, macro scale engineering, the size, shape and roughness are purely functions of manufacture process mechanisms like time of machining, depth of cut and the path, which the cutting tool adopts. Loss of co-ordination between any of these parameters leads to large errors. But in nanoscale or at molecular level, the scenario is rather strange. The size and shape are themselves so small that, roughness as a separate physical quantity is not definable.

For the sake of demarcating various engineering materials, the whole spectrum has been split into four levels.

a)  Bulk ($\mu$m): This covers all systems right from micro electro mechanical systems (MEMS) to large structures.

b)  Particles (100nm): It includes powders and composites, for example aluminium.

c)  Micro clusters (10nm): It includes alloys, catalysts, nanotubes, and fullerenes.

d)  Molecular scale (1nm):It encompasses thin films such as Langmuir-Blogett and self- assembly proteins.

Geometric features and the scale of size:

The total geometry of an object can be mathematically represented as,

$$G = S_i \bigcap S_{sh} \bigcap R$$

The sanctity in representing the geometry in such a manner is that it represents the nominal independence of the physical features. For e.g. in conventional engineering, the size ($S_i$) is much greater than shape ($S_{sh}$) and roughness (R) and is hence nominally independent of it. In other words, the values of shape and roughness do not alter the size value. In the same way, we can observe that as the size factor decreases there is growing influence of shape and roughness. As the scale reduces, it becomes progressively difficult to describe size, shape and roughness independently. This apparent arises due to difficulty in scaling down roughness.

Reduction in size is pretty simple for there is only a general reduction in all dimensions and there is not a large variation to make the reduction process so tedious. In a similar fashion, the shape is more dependent on, as already mentioned on the path, which the cutting tool takes up and is dependent on parameters that need not be rigorously balanced. To put it more perspicuously, while it does not require much approximation to measure a rectangular block, but it does requires assumption of a mean length, when the block is only a few microns in length and has rather rugged ends.

## Force Balance Variation with Scale

As the scale of size is reduced, there is profound change in the relative importance of the components of the force equations. At the mm level and above, the inertial component dominates. However, as the scale is reduced to the nanometre level, the inertial component falls off faster in magnitude than either the damping or elastic terms. The most important component becomes the elastic force. However, elastic properties are controllable since the material properties are known. The inertial term is negligible leaving the damping term which is areal and is consequently dependent on the surface finish. Adhesion and friction depend a great deal on the relatively uncontrolled surface. The change in the relative importance of the forces is the reason why insects have such different movement capabilities than mammals. Insects can change direction instantly indicating little inertia. They can walk upside down on ceilings indicating that adhesion forces are much greater than force on mass.

## Dependence of Roughness on Scale

Roughness at nanoscale can occur in three broad cases,

1) On large objects such as mirrors

2) On small objects such as in micro dynamics

3) At the molecular and atomic scale

On large objects, roughness has been regarded as irrelevant and the only probable effect it can have is on the reactivity at the surface due to increased surface area. This might probably the only limit that can be set for roughness in macro scale.

At cluster level, the priorities change. The clusters themselves are made up of few hundreds of atoms and more so, the whole functionality of the cluster lies at the surface than at the bulk. Thus what we might call roughness in macro level actually turns out to be shape at cluster level. To resolve this anomaly, the definition of roughness has been changed. While shape is defined as the configuration of molecules (that is automatically generated), the roughness is defined as the defects that arise during shape generation.

Turning our attention to the molecular level, roughness usually means the disorder in the layers of whiskers of carbon nanotubes and fibre, which could easily be avoided by careful processing and preparation techniques. Except in cases where the roughness is present at the substrate interface and the effect it causes is very much similar to what roughness might cause at macro level. Another common observation though not strictly classified, as roughness is the forming of oxides, sulphides that depends on the electron structure at the surface and hence comes under the classification of shape rather than roughness. Other such related observations are that of formation of a separate symmetry at surface of semiconductors, which produces a new unit cell at the surface called as surface reconstruction' and is a consequence of the structure seeking low potential energy. This surface reconstruction affects deposition of metal on the surface. Other phenomena include formation of surface colloids, coverage of surface with biological molecules like proteins and enzymes to impart activity etc. All these interesting applications just go on to show that roughness which is more of than not pejorative at macro scales turns out to be rather useful at nanoscale.

## Surface and Bulk Properties in Nanoscale:

Roughness in the conventional sense as we know makes little sense at nano levels. More specifically, priority should be given to how well controlled is the production process to attain the desired reactance and stability rather than on attaining smooth polished surface. The surface starts playing so important roles that, cleavage have to be done in vacuum environments to maintain reactance and stability. Exposure to atmosphere can result in adsorption of chemicals onto the surface, diffusion of these chemicals into the bulk, alteration of electric and magnetic properties and so on. This is plainly the reason why roughness was defined as the defects present on the surface.

Defects rather than roughness take over as the important surface characteristics in the nanometer and sub-nanometer domains.

Defects are usually characterized by their physical dimensions like zero-dimensional defects (point defects), one-dimensional defects (dislocations), two-dimensional defects (slip and twin) and three-dimensional defects (voids and cracks). However, the maximum dimension of any defect is purely dependent on that of the parent sample. What we call a 3-D defect at nanoscale becomes a point defect at macro scale and hence the break in definition occurs at the nanoscale.

## Engineering Shape

There are a number of shapes in engineering which are especially relevant. Of these, roundness is probably the most important. Practically every dynamic application involves rotation. The measurement and control of roundness is well understood at the macro level. However, as the size reduces it becomes more difficult, at the millimetre level and below, to include MEMS. The measurement is tricky. This is mainly due to centering. In addition, the signal as seen by the instrument is not the same as that obtained at the macro level.

For machined objects, shape is usually determined by the movement of the tool. This involves movement in two or three dimensions. The curve produced in space is usually simple, conic forms such as parabolas and spheres are often required.

## Molecular Shape and Scale of Size

The aspect of shape at the nanoscale is that it can have a meaning i.e. it is predetermined by energy or other considerations. Whereas in engineering the shape usually determines whether the work-piece can move (i.e. roundness can determine how well a shaft revolves in a bearing). At the nanoscale, the shape of an object can be a result of the internal energy balance between, say, chemical and elastic energies. In clusters or even liquid droplets, the energy balance is between Coulomb attraction and surface energy. From these considerations, it is clear that in nanotechnology shape is often an output from energy considerations rather than an output to an engineering function. Shape can therefore have completely different implications in macro- as compared with nano- regimes.

In molecular domain, shape is determined by the chemical composition and the structures have a fixed shape depending on the particular molecular configuration. In other words, the shape of a structure is preordained. The question is whether it is inhibited by defects or not. Shapes are discrete, not continuous as in macro engineering. The measurement of shape therefore is usually not the issue: failure to achieve the expected shape is.

In quality control terms in the nano- and molecular regimes shape is often an attribute and not a variable, e.g. is it the correct shape rather than the 'value' of the shape. The shape of interfaces can also influence the quantum behavior of materials. The solution of Schrodinger's electron wave equation can be influenced by the boundary conditions; an ideal parabolic shape for the potential barrier in a quantum well depends on the shape of the boundary, which could be an interface or a surface.

## Carbon Molecular Shapes

Due to their extremely precise shapes and sizes, carbon nanotubes and similar structures such as fullerenes have such remarkable electronic and mechanical properties. In principle, they can be used in gears, rack and pinion. These structures relate to the parent planar graphite.

Some enhanced mechanical properties include better stability, strength, stiffness and elastic modulus. The helicity, which is the arrangement of the carbon hexagons on their surface layer honeycomb lattice, is determined by the symmetry of the structure, the helix angle and the diameter of the tube (i.e. the shape and size).

It should be noted that it is not only carbon which has the potential for forming tubes, but also Molybdenum disulphide ($MoS_2$) the well-known lubricant also has a similar structure.

Thus to sum it up - metrological factors change in some way or the other as the size reduces below the nanometre level. The balance of forces also changes with size, with each component changing disproportionately. However, the metrological components change form rather than value. For the purely mechanical applications in microdyanmics, roughness, shape (form) and size become scrambled. Forces are small enough so that we need not consider the sub-surface stress and waviness.

Figure above shows the diversity of the geometric relationships. However there is inherent difficulty in establishing a common framework of standards to back the different types of measurements. Also, as the size decreases the nature of the signal changes. What appears as a straightforward deterministic measurement in engineering gets replaced by spatial statistical averaging and, finally, temporal statistics brought about by quantum mechanics.

## Metrology at Nanoscale

Even after monitoring carefully how various geometrical features and force balance vary in their relative meaning significance, we face yet another problem and that arises from the instrument we use.

## Stability of Signals

## Length

The system shown is conventional engineering and is stable to the signal S. Uncertainty is so small that noise is negligible. The signal is acceptable. In (b) the actual signal at any point changes with position so, to get a meaningful signal S, the geometry has to be smoothed by integration giving m1 and m2. The signal is now S = m1- m2 where m1 and m2 are mean planes. The distance S can be accurate to much less than a nanometre because of the spatial integration.

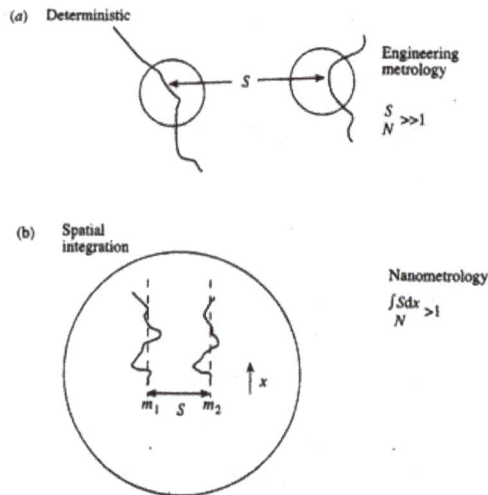

Figure: Signal form in (a) Engineering Metrology (b) Nanometrology

In (c) at molecular and atomic levels the available signals passing between the two points is limited by the distance between points p1 and p2 which are molecular or atomic un size. In this regime tunneling across the gap or through a barrier is subject to quantum laws. The only way to increase the probability of an electron or photon moving between p1 and p2 and so closing the measurement loop is to increase the observation time (i.e. to integrate temporally). This automatically makes the instrument sensitive to the environment. This temporal problem is not the same as the bandwidth of electrical noise.

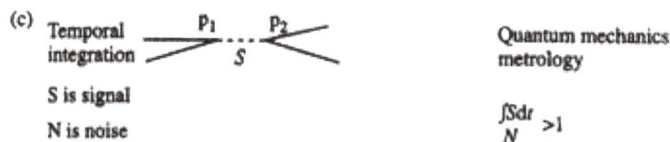

Figure: Quantum Mechanics Metrology

The increase in resolution and accuracy of modern instruments forced by application pressure has brought their specification into the nano regime-in effect almost the same scale of size spatially as that of the surface roughness. This is why the subjects of surface metrology and nanometrology are blending and why position and spacing are being lumped into the new discipline of surface nanometrology.

## Nano Position Sensing

This is defined as the technology of moving and measuring with sub-nanometre precision and has values, which are comparable to very fine surface finish. It is also obviously related to length. Optical methods can be used to measure position using white light fringes. However, a very common method is the use of piezoelectric crystals suitably energised by a known voltage to produce the movement, and some gauge, usually capacitive to control the movement. The control is necessary because, although piezoelectric crystals are very stiff, hey suffer greatly from non-linearity and hysteresis. It needs a very precise and accurate gauge to find out where the end of the crystal is in space.

As there are three components involved in the geometry of a work piece each having its idiosyncrasy it is wise to review the calibration procedure in order to see if three can be brought together in a traceable way at the nanoscale.

## Calibration

### General

The real problem lies in the way engineering surface metrology and microscopy have evolved. In many engineering applications such as those which occur in tribology have more emphasis on the height variations than on spacing or lateral dimensions. They are mainly observed for aerial structure and position in the plane of the object.

There are three types of length calibration artifacts:

1) Line spacing: In this type of calibration the effect of the shape of the probe and its interaction with the surface are small because any aberration introduced due to the interaction is common to both lines if measured from the same direction.

2) Line width

3) Size or extension

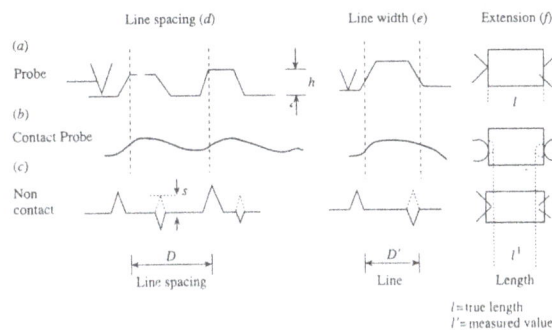

Figure: Length standards showing the effect of interaction between the probe and the artifact.

## Probe

The probe is the most important component of the new generation scanning microscopes. The process of manufacturing a tip of a particular dimension is difficult. The most common example that can be given in this case is electrochemical etching techniques to fabricate tungsten and platinum/ iridium tips for use in STM. These tips have been made with a very high aspect ratio. The main idea of developing such a tip is to obtain a very small angle at the surface. In addition to this by this process it is also possible to keep the radius of curvature very small. The technique of electrochemical etching to make STM is a natural progression from field ion microscopy. We would prefer to use platinum for tips rather than the common tungsten is due to the fact that platinum, although it is softer material, platinum is inert to oxidation. On the other hand tungsten would easily get oxidized to $WO_2$. We would also introduce some amount of iridium into the tip surface as it would tend to make the tip much harder and stiffer so that it can withstand much higher temperatures without undergoing any amount of plastic deformation.

## Some of the Unconventional Methods of Stm Tip Preparation Ion Sputter Thinning Method

The ion sputter thinning method is used to get tips which would offer a better resolution and re-liability. These tips are capable of achieving atomic resolution spatially can be fabricated. In this process the given material should be made into the dimensions in the micron level. This starting material can be obtained using the ultra-microtome cutting operation. However we should choose the appropriate material for cutting or shearing of the initial sample. Generally diamond coated tungsten carbide blades are used for this purpose. This cut sample which is obtained is then sputtered using an ion bean of Ar (Ar sputtering process). The main disadvantage of this process is that it is very slow and it would take about 18 hrs.

The novel and simple method is to develop a tip fabricated out of pencil lead by coating tungsten tips with colloidal graphite. There is also another method of tip fabrication of tips which involves the shearing of Tt/ Ir wire. By shearing the wire at a very large angle the wire gets a cut tip which is often good enough to use in measuring roughness in micro-areas.

## Use of Carbon Nanotube

One of the most novel methods is to use carbon nanotubes as a probe for STM. These would be very suitable for nanometrology. This is because the carbon nanotubes have a very small diameter, a large aspect ratio (length/ diameter). In addition to this they have the property of high stiffness to lateral movement. This very important because when we apply very large potential through the tip for rastering patterns then the tip would tend to vibrate in the direction parallel to the surface movement. This would lead to misalignment of the probe. The carbon nanotubes have high tensile strength along the axial direction which would contribute to the stiffness. There is also scope for using this carbon nano tube as a nanopipette also along with STM probe usage. Through the hollow section of the tube it is possible to send few molecules or atoms to the particular region which we have just probed. The main difficulty is that of attaching them to the body of the main pick up part.

There are also optical super tips which can be smaller than 10nm. By the use of these optical super tips is possible that only single molecules that can absorb light energy and transcribe to different optical properties. This type of tip is used in Scanning Near-Field Optical Microscope (SNOM) which is a scanning probe microscope that allows optical imaging with spatial resolution beyond the diffraction limit. A nanoscopic light source, usually a fiber tip with an aperture smaller than 100 nm, is scanned much closed to the sample surface in the region called "optical near-field". Particularly, scanning in liquids becomes easily feasible, which is vital for biological applications where the sample is always in a solution medium. For example, its applications reach from routine control of micro contact-printed samples over the determination of the orientation of proteins on surfaces to studies on single fluorescent molecules. This technique makes the sub-100 nm length scale accessible to optical investigations with all the benefits of different contrast mechanisms such as fluorescence and polarization, including the option of chemical identification. Additionally it has the benefit of recording the sample topography simultaneously with the optical information alike an atomic force microscope.

Special silicon tips capable of being used in the fabrication of semiconductor devices. For example by atomic force microscopy which involves small size movements of atom by atom, resulting in possible locations of 1014 GB/m2 on silicon wafers. The silicon tips are generally highly acicular in nature having a tip angle of only about 8-100. They are based on silicon beams and quartz tips, which are completely coated with a thin aluminum film. Despite the fact that no physical opening is created at the tip's apex prior to scanning, the energy throughput is sufficiently high and an optical resolving power of around 32 nm has been shown.

## Height Measurement Calibration at the Nanometer Scale of Size

One of the big problems associated with the height calibration at the nanometer level is that there is a tendency to try to relate the calibration procedure devised for contact methods with those of optical methods. In particular the optical methods have been used to measure the artifacts developed for stylus contact methods (note that this attempted comparison has been carried out for roughness standards. If the surface is rough the optical and mechanical probe gives the surface geometry fairly adequately. The difference between optical and mechanical techniques is not usually significant. But if the surface is smooth then even small differences can cause a serious divergence of the two methods.

Although it seems sensible to use optical interferometry to measure position, displacement and step height, typical results with interferometry where d is the measured height would give an uncertainty of measurement of $10^{-4}d$ nm + 2.5nm whereas the stylus method gives $1.5 \times 10^{-3}d$ nm + 0.7nm. The errors associated with the stylus method or nanostep method can be attributed to small non-linearities ion the transducers, whereas the errors in the optical methods are usually attributed to difficulties in fringe position measurement, surface contamination and roughness of the surface of the steps.

The true way to compare instruments properly is to measure along exactly the same track by using kinematic methods together with track identification techniques for example the double cross method as shown in the figure.

Using the relocation method there is no ambiguity of trace position. The yaw angle is adjusted until just the two marks appear on the trace. There is sometimes the possibility of occurrence of noise. This occurrence of noise may lead to the burial of the signal because there is a mismatch between what is being measured and the metrology unit. One example of this mismatch is the use of light having nominal wavelength of 0.5μm to measure nano detail. The mismatch in this case is about 500:1. This gives a very low signal to noise ratio. Also there is a limit to which what can be extracted from the attenuated signal.

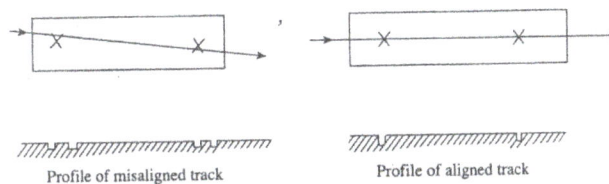

Profile of misaligned track              Profile of aligned track

Figure: Relocation method for comparing different instruments

## X-ray Interferometer

In this arrangement the interferometer is a monolithic block of silicon. This is machined to project three blades from one face. The third blade is a part of the ligament hinge. X- rays enter from the left and are focused via the second blade B onto C which is the movable blade. The movement of C varies with the absorption of X- rays by the blade. This is then detected by a scintillation counter.

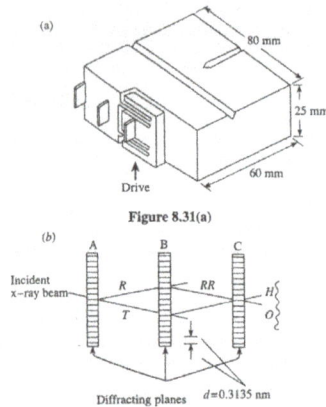

Figure: Interferometer structure (b) Arrangement of silicon blocks inside the setup.

If movable blade C is moved linearly upward the voltage output varies sinusoidally. This signal is the master which, when related back to the lattice spacing can be used to compare with the test transducer signal whose contact point or focal point is positioned on the top of the moving ligament hinge configuration of the crystal. The actual value of the x-ray wavelength is not critical but the most important factor is the positioning of the parallel faces of the blades11. Interpolation of the sinusoidal signal would give much better sub-nanometer accuracy and resolution.

## Noise

The most common source of noise in many instruments is due to electronic or thermal effects.

## Electronic

This is due to the electronic fluctuations inherent in the input stage of the amplifier or the probe itself. This noise is also commonly referred to as Johnson noise and Shot noise. The equation for RMS voltage for Johnson noise is,

$$E\left(V_j^2\right) = KT\Delta BR^{-1}$$

$\Delta B$ = bandwidth gap

K= constant which depend upon the dimensions of the tip

Shot noise is the variation in the current due to the random movement of electrons. It has a same bandwidth $\Delta B$ as above and give about the same level of equivalent noise as Johnson noise.

## Thermal Effects

The deflection of the end of an AFM cantilever beam may be caused by the Brownian motion. For a simple detection system based on viewing the end of the cantilever, the expected variation in z is given by,

$$E\left(\Delta z^{2}\right)=0.065\,K^{-0.5}$$

$$K=0.25\,E\,w h^{2}\,L^{-3}$$

Where h is thickness, w is width and L is the length of the cantilever.

In cases where the deflection is measured by a laser beam reflected from the cantilever end, an optical cantilever, then dz(L)/dx is the operative measurand.

$$Z\left(L\right)=0.667\,L\,dz\left(L\right)/\,dx$$

By applying thermal vibration theory to cover the two cases when (a) the one end of the cantilever is free and (b) when it is making contact with the surface it is possible to deduce the different modes of vibration of the beam using standard above given formulae.

When all the vibration modes are taken into account the root mean square deflection $\sqrt{z^{2}}$

is given that the optical lever is used for measurement,

$\sqrt{\left(4kt/\,r\,K\right)}$      For free end

$\sqrt{\left(kt/\,3r\,K\right)}$      For contact end

Where k is boltzman's constant and t is temperature,

For a temperature of 22 °C and a spring constant such as given by K in Nm the values of the resultant thermal noise are just less than 10-10m. This means that the mechanical noise and the electronic noise are about the same and are uncomfortably close to the desired resolution. For small structures ex small cantilevers for AFM, the internal friction has been broken down into surface effects and bulk effects, both of which dissipate energy independently.

## Calibration Artefacts

It is always very difficult to provide specimens with which to calibrate precision instruments. In engineering applications the obvious example of the use of artifacts is the use of gauge blocks, which are used as a practical intermediary between the test piece and length standards whose values can be traced back to the standard meter. However these gauge block concept cannot be extended to ranges of micrometer and nanometer.

One of the problems is the fact that there is a need for an artifact of regular pattern. This type is needed for nano-structure spacing as well as for lateral information in scanning microscopes and for filter characteristics in micro-roughness. The most common technique for measuring the roughness of the artifacts is by diffraction grating. The diamond turning is used to generate sine waves in copper. These are usually chrome plated to give hard-wearing properties[12]. However this technique cannot be used as effectively used for stylus or optical methods. This is because the stylus method tends to ignore remnant turning marks in the surface and light in optical techniques gets diffracted.

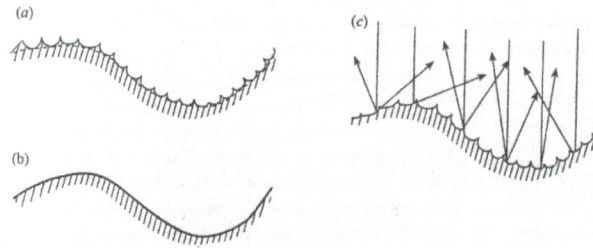

Figure: Attempt to calibrate optical instrument with tactile standard.

## Nanometrology Measurement Techniques

### Atomic Force Microscopy (AFM)

An atomic force microscope is a type of high resolution scanning probe microscope that has a resolution that you can measure in fractions of a nanometer.

It was pioneered in 1986 by Nobel Prize Winner Gerd Binnig along with Calvin Quate and Christoph Gerber.

One of the most important tools for imaging on the nanometer scale, Atomic Force Microscopy uses a cantilever with a sharp probe that scans the surface of the specimen.

When the tip of the probe travels near to a surface, the forces between the tip and sample deflect the cantilever according to Hooke's law.

Atomic force microscopy will measure a number of different forces depending on the situation and the sample that you want to measure.

As well as the forces, other microscopes can include a probe that performs more specialized measurements, such as temperature.

The force deflects the cantilever, and this changes the reflection of a laser beam that shines on the top surface of the cantilever onto an array of photodiodes. The variation of the laser beam is a measure of the applied forces.

## Contact and Non-contact Modes

There are two primary modes of operation for an atomic force microscope, namely contact mode and non-contact mode depending on whether the cantilever vibrates during the operation.

In contact mode, the cantilever drags across the sample surface and it uses the deflection of the cantilever to measure the contours of the surface.

To eliminate the noise and drift that can affect a static signal, low stiffness cantilevers are used, but this allows strong attractive forces to pull the tip to the surface. To eliminate this attraction, the tip is in contact with the surface where the overall force is repulsive.

In non-contact mode, the tip vibrates slightly above its resonance frequency and does not contact the surface of the sample. Any long range forces, like van der Waals forces, decreases the resonant frequency of the cantilever.

A feedback loop system helps to maintain the oscillation amplitude constant by changing the distance from the tip to the sample. Recording the distance between the tip and sample at each point allows the software to construct a topographic image of the sample surface.

Most samples will form a layer of moisture on the surface if stored at ambient conditions, and this can make it difficult to measure the sample accurately.

If the probe tip is close enough to detect the short-range forces then it is close enough to stick to the moisture. One way around this is tapping, or dynamic contact mode.

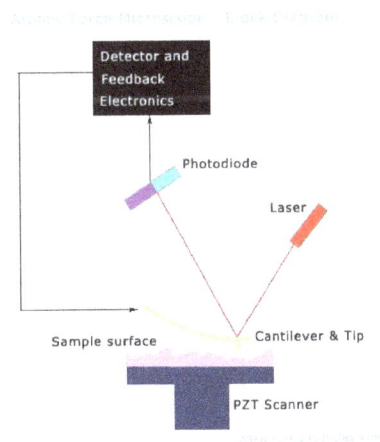

## Tapping Mode

In tapping mode, the cantilever uses a piezoelectric element mounted on the top to oscillate it at near to its resonance frequency with an amplitude of up to 200nm.

The forces cause the amplitude to decrease as the tip gets close to the surface, and the height of the cantilever adjusts to keep the amplitude constant.

This tapping results in less damage to the sample than contact mode and is more accurate than non-contact mode when moisture is present on a sample.

## Topographic Image

Image formation is a plotting method that produces a color mapping through changing the x-y position of the tip while scanning and recording the measured variable, i.e. the intensity of control signal, to each x-y coordinate. The color mapping shows the measured value corresponding to each coordinate. The image expresses the intensity of a value as a hue. Usually, the correspondence between the intensity of a value and a hue is shown as a color scale in the explanatory notes accompanying the image.

## What is the Topographic Image of Atomic Force Microscope

Operation mode of image forming of the AFM are generally classified into two groups from the viewpoint whether it uses z-Feedback loop (not shown) to maintain the tip-sample distance to keep signal intensity exported by the detector. The first one (using z-Feedback loop), said to be "constant **XX** mode" (**XX** is something which kept by z-Feedback loop).

Topographic image formation mode is based on abovementioned "constant **XX** mode", z-Feedback loop controls the relative distance between the probe and the sample through outputting control signals to keep constant one of frequency, vibration and phase which typically corresponds to the motion of cantilever (for instance, voltage is applied to the Z-piezoelectric element and it moves the sample up and down towards the Z direction.

Details will be explained in the case that especially "constant df mode"(FM-AFM) among AFM as an instance in next section.

## Topographic Image of FM-AFM

When the distance between the probe and the sample is brought to the range where atomic force may be detected, while a cantilever is excited in its natural eigen frequency ($f_o$), a phenomenon occurs that the resonance frequency (f) of the cantilever shifts from its original resonance frequency (natural eigen frequency). In other words, in the range where atomic force may be detected, the frequency shift ($df=f-f_o$) will be observed. So, when the distance between the probe and the sample is in the non-contact region, the frequency shift increases in negative direction as the distance between the probe and the sample gets smaller.

When the sample has concavity and convexity, the distance between the tip-apex and the sample varies in accordance with the concavity and convexity accompanied with a scan of the sample along x-y direction (without height regulation in z-direction). As a result, the frequency shift arises. The image in which the values of the frequency obtained by a raster scan along the x-y direction of the sample surface are plotted against the x-y coordination of each measurement point is called a constant-height image.

On the other hand, the df may be kept constant by moving the probe upward and downward in z-direction using a negative feedback (by using z-feedback loop) while the raster scan of the sample surface along the x-y direction. The image in which the amounts of the negative feedback (the moving distance of the probe upward and downward in z-direction) are plotted against the x-y coordination of each measurement point is a topographic image. In other words, the topographic image is a trace of the tip of the probe regulated so that the df is constant and it may also be considered to be a plot of a constant-height surface of the df.

Therefore, the topographic image of the AFM is not the exact surface morphology itself, but actually the image influenced by the bond-order between the probe and the sample, however, the topographic image of the AFM is considered to reflect the geographical shape of the surface more than the topographic image of a scanning tunnel microscope.

## Force Spectroscopy

Another major application of AFM (besides imaging) is force spectroscopy, the direct measurement of tip-sample interaction forces as a function of the gap between the tip and sample (the result of this measurement is called a force-distance curve). For this method, the AFM tip is extended towards and retracted from the surface as the deflection of the cantilever is monitored as a function of piezoelectric displacement. These measurements have been used to measure nanoscale contacts, atomic bonding, Van der Waals forces, and Casimir forces, dissolution forces in liquids and single molecule stretching and rupture forces. Furthermore, AFM was used to measure, in an aqueous environment, the dispersion force due to polymer adsorbed on the substrate. Forces of the order of a few piconewtons can now be routinely measured with a vertical distance resolution of better than 0.1 nanometers. Force spectroscopy can be performed with either static or dynamic modes. In dynamic modes, information about the cantilever vibration is monitored in addition to the static deflection.

Problems with the technique include no direct measurement of the tip-sample separation and the common need for low-stiffness cantilevers, which tend to 'snap' to the surface. These problems are not insurmountable. An AFM that directly measures the tip-sample separation has been developed. The snap-in can be reduced by measuring in liquids or by using stiffer cantilevers, but in the latter case a more sensitive deflection sensor is needed. By applying a small dither to the tip, the stiffness (force gradient) of the bond can be measured as well.

## Biological Applications and other

Force spectroscopy is used in biophysics to measure the mechanical properties. of living material (such as tissue or cells) or detect structures of different stiffness buried into the bulk of the sample using the stiffness tomography. Another application was to measure the interaction forces between from one hand a material stuck on the tip of the cantilever, and from another hand the surface of particles either free or occupied by the same material. From the adhesion force distribution curve, a mean value of the forces has been derived. It allowed to make a cartography of the surface of the particles, covered or not by the material.

### Identification of Individual Surface Atoms

The AFM can be used to image and manipulate atoms and structures on a variety of surfaces. The atom at the apex of the tip "senses" individual atoms on the underlying surface when it forms incipient chemical bonds with each atom. Because these chemical interactions subtly alter the tip's vibration frequency, they can be detected and mapped. This principle was used to distinguish between atoms of silicon, tin and lead on an alloy surface, by comparing these 'atomic fingerprints' to values obtained from large-scale density functional theory (DFT) simulations.

The trick is to first measure these forces precisely for each type of atom expected in the sample, and then to compare with forces given by DFT simulations. The team found that the tip interacted most strongly with silicon atoms, and interacted 24% and 41% less strongly with tin and lead atoms, respectively. Thus, each different type of atom can be identified in the matrix as the tip is moved across the surface.

## Probe

An AFM probe has a sharp tip on the free-swinging end of a cantilever that is protruding from a holder. The dimensions of the cantilever are in the scale of micrometers. The radius of the tip is usually on the scale of a few nanometers to a few tens of nanometers. (Specialized probes exist with much larger end radii, for example probes for indentation of soft materials.) The cantilever holder, also called holder chip – often 1.6 mm by 3.4 mm in size – allows the operator to hold the AFM cantilever/probe assembly with tweezers and fit it into the corresponding holder clips on the scanning head of the atomic force microscope.

This device is most commonly called an "AFM probe", but other names include "AFM tip" and "cantilever" (employing the name of a single part as the name of the whole device). An AFM probe is a particular type of SPM (scanning probe microscopy) probe.

AFM probes are manufactured with MEMS technology. Most AFM probes used are made from silicon (Si), but borosilicate glass and silicon nitride are also in use. AFM probes are considered consumables as they are often replaced when the tip apex becomes dull or contaminated or when the cantilever is broken. They can cost from a couple of tens of dollars up to hundreds of dollars per cantilever for the most specialized cantilever/probe combinations.

Just the tip is brought very close to the surface of the object under investigation, the cantilever is deflected by the interaction between the tip and the surface, which is what the AFM is designed to measure. A spatial map of the interaction can be made by measuring the deflection at many points on a 2D surface.

Several types of interaction can be detected. Depending on the interaction under investigation, the surface of the tip of the AFM probe needs to be modified with a coating. Among the coatings used are gold – for covalent bonding of biological molecules and the detection of their interaction with a surface, diamond for increased wear resistance and magnetic coatings for detecting the magnetic properties of the investigated surface. Another solution exists to achieve high resolution magnetic imaging : having the probe equip with a microSQUID. The AFM tips is fabricated using silicon micro machining and the precise positioning of the microSQUID loop is done by electron beam lithography.

The surface of the cantilevers can also be modified. These coatings are mostly applied in order to increase the reflectance of the cantilever and to improve the deflection signal.

## Forces vs Tip Geometry

The forces between the tip and the sample strongly depend on the geometry of the tip. Various studies were exploited in the past years to write the forces as a function of the tip parameters.

Among the different forces between the tip and the sample, the water meniscus forces are highly interesting, both in air and liquid environment. Other forces must be considered, like the Coulomb force, van der Waals forces, double layer interactions, solvation forces, hydration and hydrophobic forces.

## Water Meniscus

Water meniscus forces are highly interesting for AFM measurements in air. Due to the ambient humidity, a thin layer of water is formed between the tip and the sample during air measumements. The resulting capillary force gives rise to a strong attractive force that pulls the tip onto the surface. In fact, the adhesion force measured between tip and sample in ambient air of finite humidity is usually dominated by capillary forces. As a consequence, it is difficult to pull the tip away from the surface. For soft samples including many polymers and in particular biological materials, the strong adhesive capillary force gives rise to sample degradation and distruction upon imaging in contact mode. Historically, these problems were an important motivation for the development of dynamic imaging in air (e.g. 'tapping mode'). During tapping mode imaging in air, capillary bridges still form. Yet, for suitable imaging conditions, the capillary bridges are formed and broken in every oscillation cycle of the cantilever normal to the surface, as can be inferred from a analysis of cantilever amplitude and phase vs. distance curves. As a consequence, distructive shear forces are largely reduced and soft samples can be investigated.

In order to quantify the equilibrium capillary force, it is necessary to start from the Laplace equation for pressure:

$$P = \gamma_L \left( \frac{1}{r_1} + \frac{1}{r_0} \right) \simeq \frac{\gamma_L}{r_{eff}}$$

where $\gamma_L$ is the surface energy and $r_0$ and $r_1$ are defined in the figure.

The pressure is applied on an area of,

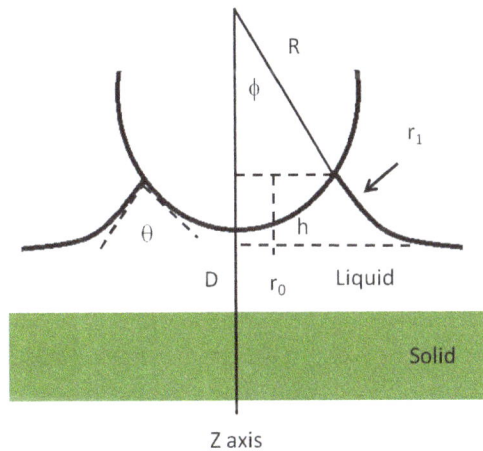

Model for AFM water meniscus.

$$A \simeq 2\pi R \simeq \left[ reff \left( 1 + \cos\theta \right) + h \right]$$

where d, $\theta$, and h are defined in the figure.

The force which pulles together the two surfaces is,

$$F = 2\pi R \gamma_L \left( 1 + \cos\theta + \frac{h}{r_{eff}} \right)$$

The same formula could also be calculated as a function of relative humidity.

Gao calculated formulas for different tip geometries. As an example, the forse decreases by 20% for a conical tip with respect to a spherical tip.

When these forces are calculated, a difference must be made between the wet on dry situation and the wet on wet situation.

For a spherical tip, the force is:

$$f_m = -2\pi R\gamma_L \left(\cos\theta + \cos\phi\right)\left(1 - \frac{dh}{dD}\right)$$

for dry on wet

$$f_m = -2\pi R\gamma_L \frac{dr_0}{dD}$$

for wet on wet

where $\theta$ is the contact angle of the dry sphere and $\phi$ is the immersed angle, as shown in the figure Also R,h and D are illustrated in the same figure.

For a conical tip, the formula becomes:

$$f_m = -2\pi R\gamma_L \frac{\tan\delta}{\cos\delta}\left(\cos\theta + \sin\delta\right)\left(hD\right)\left(1 - \frac{dh}{dD}\right)$$

for dry on wet

$$f_m = -2\pi R\gamma_L \left(\frac{1}{\cos\delta} + \sin\delta\right)\left(r_0\right)\left(\frac{dr_0}{dD}\right)$$

for wet on wet

where $\delta$ is the half cone angle and $r_0$ and h are parameters of the meniscus profile.

## AFM Cantilever-deflection Measurement

## Beam-deflection Measurement

The most common method for cantilever-deflection measurements is the beam-deflection method. In this method, laser light from a solid-state diode is reflected off the back of the cantilever and collected by a position-sensitive detector (PSD) consisting of two closely spaced photodiodes, whose output signal is collected by a differential amplifier. Angular displacement of the cantilever results in one photodiode collecting more light than the other photodiode, producing an output signal (the difference between the photodiode signals normalized by their sum), which is proportional to the deflection of the cantilever. The sensitivity of the beam-deflection method is very high, a noise floor on the order of 10 fm $Hz^{-1/2}$ can be obtained routinely in a well-designed system. Although this method is sometimes called the 'optical lever' method, the signal is not amplified if the beam path is made longer. A longer beam path increases the motion of the reflected spot on the photodiodes, but also widens the spot by the same amount due to diffraction, so that the same amount of optical power is moved from one photodiode to the other. The 'optical leverage' (output signal of the detector divided by deflection of the cantilever) is inversely proportional to the numerical aperture of the beam focusing optics, as long as the focused laser spot is small enough to fall completely on the cantilever. It is also inversely proportional to the length of the cantilever.

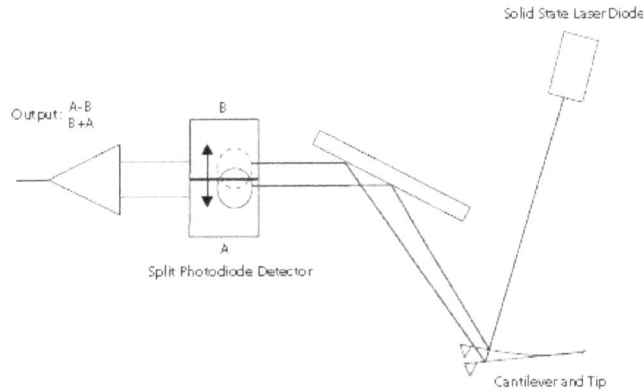

AFM beam-deflection detection

The relative popularity of the beam-deflection method can be explained by its high sensitivity and simple operation, and by the fact that cantilevers do not require electrical contacts or other special treatments, and can therefore be fabricated relatively cheaply with sharp integrated tips.

## Other Deflection-measurement Methods

Many other methods for beam-deflection measurements exist.

- Piezoelectric detection: Cantilevers made from quartz (such as the qPlus configuration), or other piezoelectric materials can directly detect deflection as an electrical signal. Cantilever oscillations down to 10pm have been detected with this method.

- Laser Doppler vibrometry: A laser Doppler vibrometer can be used to produce very accurate deflection measurements for an oscillating cantilever (thus is only used in non-contact mode). This method is expensive and is only used by relatively few groups.

- Scanning tunneling microscope (STM): The first atomic microscope used an STM complete with its own feedback mechanism to measure deflection. This method is very difficult to implement, and is slow to react to deflection changes compared to modern methods.

- Optical interferometry: Optical interferometry can be used to measure cantilever deflection. Due to the nanometre scale deflections measured in AFM, the interferometer is running in the sub-fringe regime, thus, any drift in laser power or wavelength has strong effects on the measurement. For these reasons optical interferometer measurements must be done with great care (for example using index matching fluids between optical fibre junctions), with very stable lasers. For these reasons optical interferometry is rarely used.

- Capacitive detection: Metal coated cantilevers can form a capacitor with another contact located behind the cantilever. Deflection changes the distance between the contacts and can be measured as a change in capacitance.

- Piezoresistive detection: Cantilevers can be fabricated with piezoresistive elements that act as a strain gauge. Using a Wheatstone bridge, strain in the AFM cantilever due to deflection can be measured. This is not commonly used in vacuum applications, as the piezoresistive detection dissipates energy from the system affecting Q of the resonance.

## Piezoelectric Scanners

AFM scanners are made from piezoelectric material, which expands and contracts proportionally to an applied voltage. Whether they elongate or contract depends upon the polarity of the voltage applied. Traditionally the tip or sample is mounted on a 'tripod' of three piezo crystals, with each responsible for scanning in the $x,y$ and $z$ directions. In 1986, the same year as the AFM was invented, a new piezoelectric scanner, the tube scanner, was developed for use in STM. Later tube scanners were incorporated into AFMs. The tube scanner can move the sample in the $x$, $y$, and $z$ directions using a single tube piezo with a single interior contact and four external contacts. An advantage of the tube scanner compared to the original tripod design, is better vibrational isolation, resulting from the higher resonant frequency of the single element construction, in combination with a low resonant frequency isolation stage. A disadvantage is that the $x$-$y$ motion can cause unwanted $z$ motion resulting in distortion. Another popular design for AFM scanners is the flexure stage, which uses separate piezos for each axis, and couples them through a flexure mechanism.

Scanners are characterized by their sensitivity, which is the ratio of piezo movement to piezo voltage, i.e., by how much the piezo material extends or contracts per applied volt. Because of differences in material or size, the sensitivity varies from scanner to scanner. Sensitivity varies non-linearly with respect to scan size. Piezo scanners exhibit more sensitivity at the end than at the beginning of a scan. This causes the forward and reverse scans to behave differently and display hysteresis between the two scan directions. This can be corrected by applying a non-linear voltage to the piezo electrodes to cause linear scanner movement and calibrating the scanner accordingly. One disadvantage of this approach is that it requires re-calibration because the precise non-linear voltage needed to correct non-linear movement will change as the piezo ages. This problem can be circumvented by adding a linear sensor to the sample stage or piezo stage to detect the true movement of the piezo. Deviations from ideal movement can be detected by the sensor and corrections applied to the piezo drive signal to correct for non-linear piezo movement. This design is known as a 'closed loop' AFM. Non-sensored piezo AFMs are referred to as 'open loop' AFMs.

The sensitivity of piezoelectric materials decreases exponentially with time. This causes most of the change in sensitivity to occur in the initial stages of the scanner's life. Piezoelectric scanners are run for approximately 48 hours before they are shipped from the factory so that they are past the point where they may have large changes in sensitivity. As the scanner ages, the sensitivity will change less with time and the scanner would seldom require recalibration, though various manufacturer manuals recommend monthly to semi-monthly calibration of open loop AFMs.

## Advantages and Disadvantages

The first atomic force microscope

## Advantages

AFM has several advantages over the scanning electron microscope (SEM). Unlike the electron microscope, which provides a two-dimensional projection or a two-dimensional image of a sample, the AFM provides a three-dimensional surface profile. In addition, samples viewed by AFM do not require any special treatments (such as metal/carbon coatings) that would irreversibly change or damage the sample, and does not typically suffer from charging artifacts in the final image. While an electron microscope needs an expensive vacuum environment for proper operation, most AFM modes can work perfectly well in ambient air or even a liquid environment. This makes it possible to study biological macromolecules and even living organisms. In principle, AFM can provide higher resolution than SEM. It has been shown to give true atomic resolution in ultra-high vacuum (UHV) and, more recently, in liquid environments. High resolution AFM is comparable in resolution to scanning tunneling microscopy and transmission electron microscopy. AFM can also be combined with a variety of optical microscopy and spectroscopy techniques such as fluorescent microscopy of infrared spectroscopy, giving rise to scanning near-field optical microscopy, nano-FTIR and further expanding its applicability. Combined AFM-optical instruments have been applied primarily in the biological sciences but have recently attracted strong interest in photovoltaics and energy-storage research, polymer sciences, nanotechnology and even medical research.

## Disadvantages

A disadvantage of AFM compared with the scanning electron microscope (SEM) is the single scan image size. In one pass, the SEM can image an area on the order of square millimeters with a depth of field on the order of millimeters, whereas the AFM can only image a maximum scanning area of about 150×150 micrometers and a maximum height on the order of 10-20 micrometers. One method of improving the scanned area size for AFM is by using parallel probes in a fashion similar to that of millipede data storage.

The scanning speed of an AFM is also a limitation. Traditionally, an AFM cannot scan images as fast as an SEM, requiring several minutes for a typical scan, while an SEM is capable of scanning at near real-time, although at relatively low quality. The relatively slow rate of scanning during AFM imaging often leads to thermal drift in the image making the AFM less suited for measuring accurate distances between topographical features on the image. However, several fast-acting designs  were suggested to increase microscope scanning productivity including what is being termed videoAFM (reasonable quality images are being obtained with videoAFM at video rate: faster than the average SEM). To eliminate image distortions induced by thermal drift, several methods have been introduced.

AFM images can also be affected by nonlinearity, hysteresis, and creep of the piezoelectric material and cross-talk between the $x$, $y$, $z$ axes that may require software enhancement and filtering. Such filtering could "flatten" out real topographical features. However, newer AFMs utilize real-time correction software (for example, feature-oriented scanning) or closed-loop scanners, which practically eliminate these problems. Some AFMs also use separated orthogonal scanners (as opposed to a single tube), which also serve to eliminate part of the cross-talk problems.

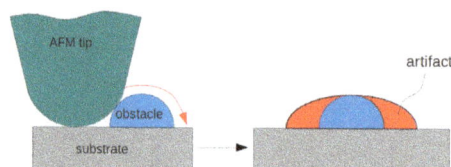

Showing an AFM artifact arising from a tip with a high radius of curvature with respect to the feature that is to be visualized.

As with any other imaging technique, there is the possibility of image artifacts, which could be induced by an unsuitable tip, a poor operating environment, or even by the sample itself, as depicted on the right. These image artifacts are unavoidable; however, their occurrence and effect on results can be reduced through various methods. Artifacts resulting from a too-coarse tip can be caused for example by inappropriate handling or de facto collisions with the sample by either scanning too fast or having an unreasonably rough surface, causing actual wearing of the tip.

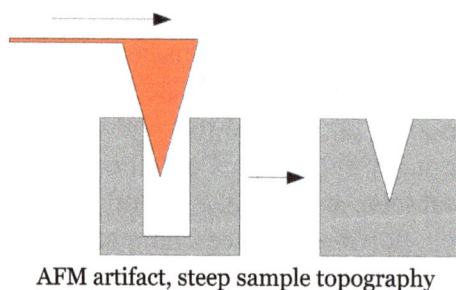

AFM artifact, steep sample topography

Due to the nature of AFM probes, they cannot normally measure steep walls or overhangs. Specially made cantilevers and AFMs can be used to modulate the probe sideways as well as up and down (as with dynamic contact and non-contact modes) to measure sidewalls, at the cost of more expensive cantilevers, lower lateral resolution and additional artifacts.

## Other Applications in Various Fields of Study

The latest efforts in integrating nanotechnology and biological research have been successful and show much promise for the future. Since nanoparticles are a potential vehicle of drug delivery, the biological responses of cells to these nanoparticles are continuously being explored to optimize their efficacy and how their design could be improved. Pyrgiotakis et al. were able to study the interaction between $CeO_2$ and $Fe_2O_3$ engineered nanoparticles and cells by attaching the engineered nanoparticles to the AFM tip. Studies have taken advantage of AFM to obtain further information on the behavior of live cells in biological media. Real-time atomic force spectroscopy (or nanoscopy) and dynamic atomic force spectroscopy have been used to study live cells and membrane proteins and their dynamic behavior at high resolution, on the nanoscale. Imaging and obtaining information on the topography and the properties of the cells has also given insight into chemical processes and mechanisms that occur through cell-cell interaction and interactions with other signaling molecules (ex. ligands). Evans and Calderwood used single cell force microscopy to study cell adhesion forces, bond kinetics/dynamic bond strength and its role in chemical processes such as cell signaling. Scheuring, Lévy, and Rigaud reviewed studies in which AFM to explore the crystal structure of membrane proteins of photosynthetic bacteria.Alsteen et al. have used AFM-based nanoscopy to perform a real-time analysis of the interaction between live mycobacteria and antimycobacterial drugs (specifically isoniazid, ethionamide, ethambutol, and streptomycine), which serves as an example of the more in-depth analysis of pathogen-drug interactions that can be done through AFM.

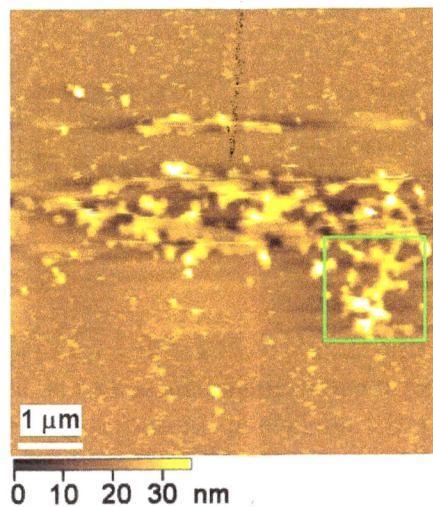

AFM image of part of a Golgi apparatus isolated from HeLa cells

## Scanning Tunneling Microscopy (STM)

Scanning tunneling microscope (STM) is a type of microscope whose principle of operation is based on the quantum mechanical phenomenon known as tunneling, in which the wavelike properties of electrons permit them to "tunnel" beyond the surface of a solid into regions of space that are forbidden to them under the rules of classical physics. The probability of finding such tunneling electrons decreases exponentially as the distance from the surface increases. The STM makes use of this extreme sensitivity to distance. The sharp tip of a tungsten needle is positioned a few angstroms from the sample surface. A small voltage is applied between the probe tip and the surface, causing electrons to tunnel across the gap. As the probe is scanned over the surface, it registers variations in the tunneling current, and this information can be processed to provide a topographical image of the surface.

The STM appeared in 1981, when Swiss physicists Gerd Binnig and Heinrich Rohrer set out to build a tool for studying the local conductivity of surfaces. Binnig and Rohrer chose the surface of gold for their first image. When the image was displayed on the screen of a television monitor, they saw rows of precisely spaced atoms and observed broad terraces separated by steps one atom in height. Binnig and Rohrer had discovered in the STM a simple method for creating a direct image of the atomic structure of surfaces. Their discovery opened a new era for surface science, and their impressive achievement was recognized with the award of the Nobel Prize for Physics in 1986.

## The Quantum Corral

The STM image below shows the direction of standing-wave patterns in the local density of states of the Cu (111) surface. These spatial oscillations are quantum-mechanical interference patterns caused by scattering of the two-dimensional electron gas off the Fe atoms and point defects.

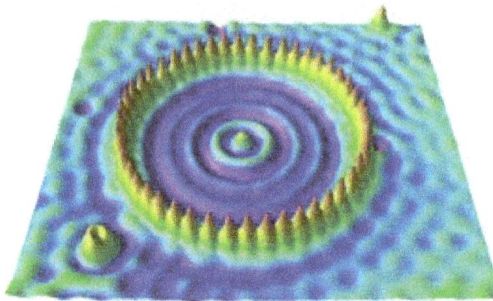

Courtesy of International Business Machines Corporation. Unauthorized use not permitted

## How an STM Works

The scanning tunneling microscope (STM) works by scanning a very sharp metal wire tip over a surface. By bringing the tip very close to the surface, and by applying an electrical voltage to the tip or sample, we can image the surface at an extremely small scale – down to resolving individual atoms.

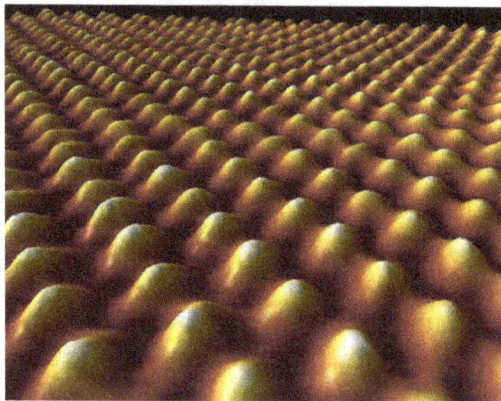

3D rendered Scanning Tunneling Microscope image of atoms.

The STM is based on several principles. One is the quantum mechanical effect of tunneling. It is this effect that allows us to "see" the surface. Another principle is the piezoelectric effect. It is this effect that allows us to precisely scan the tip with angstrom-level control. Lastly, a feedback loop is required, which monitors the tunneling current and coordinates the current and the positioning of the tip. This is shown schematically below where the tunneling is from tip to surface with the tip rastering with piezoelectric positioning, with the feedback loop maintaining a current setpoint to generate a 3D image of the electronic topography:

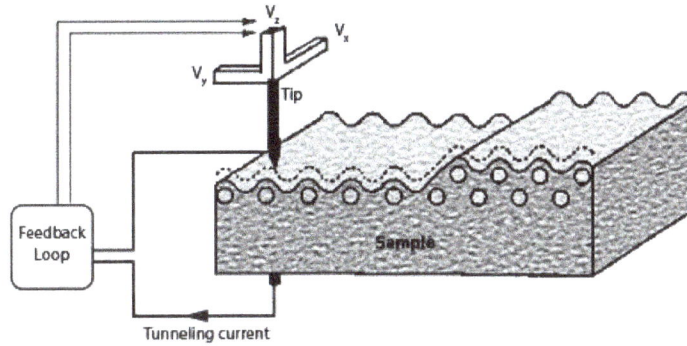

Schematic of scanning tunneling microscopy (STM).

# Tunneling

Tunneling is a functioning concept that arises from quantum mechanics. Classically, an object hitting an impenetrable barrier will not pass through. In contrast, objects with a very small mass, such as the electron, have wavelike characteristics which permit such an event, referred to as tunneling.

Electrons behave as beams of energy, and in the presence of a potential $U(z)$, assuming 1-dimensional case, the energy levels $\psi_n(z)$ of the electrons are given by solutions to Schrödinger's equation,

$$-\frac{\hbar^2}{2m}\frac{\partial^2\psi_n(z)}{\partial z^2}+U(z)\psi_n(z)=E\psi_n(z)$$

where $\hbar$ is the reduced Planck's constant, $z$ is the position, and $m$ is the mass of an electron. If an electron of energy $E$ is incident upon an energy barrier of height $U(z)$, the electron wave function is a traveling wave solution,

$$\psi_n(z)=\psi_n(0)e^{\pm ikz}$$

Where,

$$k=\frac{\sqrt{2m(E-U(z))}}{\hbar}$$

if E > U(z), which is true for a wave function inside the tip or inside the sample. Inside a barrier, E < U(z) so the wave functions which satisfy this are decaying waves,

$$\psi_n(z)=\psi_n(0)e^{\pm\kappa z}$$

Where,

$$\kappa=\frac{\sqrt{2m(U-E)}}{\hbar}$$

uantifies the decay of the wave inside the barrier, with the barrier in the +z direction for $-\kappa$.

| Condensed matter experiments |
|---|
|  |
| ARPES |
| ACAR |
| Neutron scattering |
| X-ray spectroscopy |
| Quantum oscillations |
| Scanning tunneling microscopy |

Knowing the wave function allows one to calculate the probability density for that electron to be found at some location. In the case of tunneling, the tip and sample wave functions overlap such that when under a bias, there is some finite probability to find the electron in the barrier region and even on the other side of the barrier. Let us assume the bias is $V$ and the barrier width is $W$. This probability, $P$, that an electron at $z=0$ (left edge of barrier) can be found at $z=W$ (right edge of barrier) is proportional to the wave function squared,

$$P \propto |\psi_n(0)|^2 \, e^{-2\kappa W}.$$

If the bias is small, we can let $U - E \approx \varphi M$ in the expression for $\kappa$, where $\varphi M$, the work function, gives the minimum energy needed to bring an electron from an occupied level, the highest of which is at the Fermi level (for metals at $T=0$ kelvins), to vacuum level. When a small bias $V$ is applied to the system, only electronic states very near the Fermi level, within $eV$ (a product of electron charge and voltage, not to be confused here with electronvolt unit), are excited. These excited electrons can tunnel across the barrier. In other words, tunneling occurs mainly with electrons of energies near the Fermi level.

A large scanning tunneling microscope, in the labs of the London Centre for Nanotechnology

However, tunneling does require that there be an empty level of the same energy as the electron for the electron to tunnel into on the other side of the barrier. It is because of this restriction that the tunneling current can be related to the density of available or filled states in the sample. The current due to an applied voltage $V$ (assume tunneling occurs sample to tip) depends on two factors: 1) the number of electrons between $E_f$ and $eV$ in the sample, and 2) the number among them which have corresponding free states to tunnel into on the other side of the barrier at the tip. The higher the density of available states the greater the tunneling current. When $V$ is positive, electrons in the tip tunnel into empty states in the sample; for a negative bias, electrons tunnel out of occupied states in the sample into the tip.

Mathematically, this tunneling current is given by,

$$I \propto \sum_{E_f - eV}^{E_f} |\psi_n(0)|^2 \, e^{-2\kappa W}.$$

One can sum the probability over energies between $E_f - eV$ and $E_f$ to get the number of states available in this energy range per unit volume, thereby finding the local density of states (LDOS) near the Fermi level. The LDOS near some energy $E$ in an interval $\varepsilon$ is given by,

$$\rho_s(z, E) = \frac{1}{\epsilon} \sum_{E-\epsilon}^{E} |\psi_n(z)|^2,$$

and the tunnel current at a small bias V is proportional to the LDOS near the Fermi level, which gives important information about the sample. It is desirable to use LDOS to express the current because this value does not change as the volume changes, while probability density does. Thus the tunneling current is given by,

$$I \propto V \rho_s(0, E_f) e^{-2\kappa W}$$

where $\rho_s(0, E_f)$ is the LDOS near the Fermi level of the sample at the sample surface. This current can also be expressed in terms of the LDOS near the Fermi level of the sample at the tip surface,

$$I \propto V \, {}_s(W, E_f)$$

The exponential term in the above equations means that small variations in W greatly influence the tunnel current. If the separation is decreased by 1 Å, the current increases by an order of magnitude, and vice versa.

This approach fails to account for the *rate* at which electrons can pass the barrier. This rate should affect the tunnel current, so it can be treated using the Fermi's golden rule with the appropriate tunneling matrix element. John Bardeen solved this problem in his study of the metal-insulator-metal junction. He found that if he solved Schrödinger's equation for each side of the junction separately to obtain the wave functions ψ and χ for each electrode, he could obtain the tunnel matrix, M, from the overlap of these two wave functions.This can be applied to STM by making the electrodes the tip and sample, assigning ψ and χ as sample and tip wave functions, respectively, and evaluating M at some surface S between the metal electrodes, where z=0 at the sample surface and z=W at the tip surface.

Now, Fermi's Golden Rule gives the rate for electron transfer across the barrier, and is written

$$w = \frac{2\pi}{\hbar} |M|^2 \, \delta(E_\psi - E_\chi),$$

where $\delta(E_\psi - E_\chi)$ restricts tunneling to occur only between electron levels with the same energy. The tunnel matrix element, given by,

$$M = \frac{\hbar^2}{2m} \int_{z=z_0} (\chi^* \frac{\partial \psi}{\partial z} - \psi \frac{\partial \chi^*}{\partial z}) dS,$$

is a description of the lower energy associated with the interaction of wave functions at the overlap, also called the resonance energy.

Summing over all the states gives the tunneling current as,

$$I = \frac{4\pi e}{\hbar} \int_{-\infty}^{+\infty} \left[ f\left(E_f - eV + e\right) - f\left(E_f + e\right) \right] \rho_s \left(E_f - eV + e\right) \rho_T \left(E_f + e\right) |M|^2 \, de$$

where $f$ is the Fermi function, $\rho_s$ and $\rho_T$ are the density of states in the sample and tip, respectively. The Fermi distribution function describes the filling of electron levels at a given temperature T.

## Piezoelectric Effect

The piezoelectric effect was discovered by Pierre Curie in 1880. The effect is created by squeezing the sides of certain crystals, such as quartz or barium titanate. The result is the creation of opposite charges on the sides. The effect can be reversed as well; by applying a voltage across a piezoelectric crystal, it will elongate or compress.

These materials are used to scan the tip in an scanning tunneling microscopy (STM) and most other scanning probe techniques. A typical piezoelectric material used in scanning probe microscopy is PZT (lead zirconium titanate).

## Feedback Loop

Electronics are needed to measure the current, scan the tip, and translate this information into a form that we can use for STM imaging. A feedback loop constantly monitors the tunneling current and makes adjustments to the tip to maintain a constant tunneling current. These adjustments are recorded by the computer and presented as an image in the STM software. Such a setup is called a constant current image.

In addition, for very flat surfaces, the feedback loop can be turned off and only the current is displayed. This is a constant height image.

## Applications

Several surfaces have been studied with the STM. The arrangement of individual atoms on the metal surfaces of gold, platinum, nickel, and copper have all been accurately documented. The absorption and diffusion of different species such as oxygen and the epitaxial growth of gold on gold, silver on gold, and nickel on gold also have been examined in detail.

The surfaces of silicon have been studied more extensively than those of any other material. The surfaces are prepared by being heated in vacuum to temperatures so high that the atoms there rearrange their positions in a process called surface reconstruction. The reconstruction of the silicon surface designated (111) has been studied in minute detail. Such a surface reconstructs into an intricate and complex pattern known as the Takayanagi $7 \times 7$ structure. The position, the chemical reactivity, and the electronic configuration of each atomic site on the $7 \times 7$ surface has been measured with the STM. The reconstruction of the silicon surface designated (100) is more simple. The surface atoms form pairs, or dimers, that fit into rows that extend across the entire silicon surface.

"Vacuum tunneling" of electrons from tip to sample can take place even though the environment in the region surrounding the tip is not a vacuum but is filled with molecules of gas or liquids. With a tip-sample spacing as small as five angstroms, there is little room for molecules—even though they may exist in the surrounding atmosphere. The STM can operate in ambient atmosphere as well as in high vacuum. Indeed, it has been operated in air, in water, in insulating fluids, and in the ionic solutions used in electrochemistry. It is much more convenient than ultrahigh-vacuum instruments. When a high-vacuum environment is employed, its purpose is not to improve the performance of the STM but rather to ensure the cleanliness of the sample surface.

The STM can be cooled to temperatures less than 4 K (–269 °C, or –452 °F)—the temperature of liquid helium. It can be heated above 973 K (700 °C, or 1,300 °F). The low temperature is used to investigate the properties of superconducting materials, while the high temperature is employed to study the rapid diffusion of atoms across the surface of metals and their corrosion.

The STM is used primarily for imaging, but there are many other modalities that have been explored. The strong electric field between tip and sample has been utilized to move atoms along the sample surface. It has been used to enhance the etching rates in various gases. In one instance, a voltage of four volts was applied; the field at the tip was strong enough to remove atoms from the tip and deposit them on a substrate. This procedure has been employed with a gold tip to fabricate small gold islands or clusters on the substrate with several hundred atoms of gold in each cluster. These nanostructures are used to pattern the surface on a scale that is unprecedented.

## Scanning Electron Microscopy (SEM)

A scanning electron microscope (SEM) scans a focused electron beam over a surface to create an image. The electrons in the beam interact with the sample, producing various signals that can be used to obtain information about the surface topography and composition

## Why use Electrons Instead of Light in a Microscope

Given sufficient light, the human eye can distinguish two points 0.2 mm apart, without the aid of any additional lenses. This distance is called the resolving power or resolution of the eye. A lens or an assembly of lenses (a microscope) can be used to magnify this distance and enable the eye to see points even closer together than 0.2 mm.

A modern light microscope has a maximum magnification of about 1000x. The resolving power of the microscope was not only limited by the number and quality of the lenses but also by the wavelength of the light used for illumination. White light has wavelengths from 400 to 700 nanometers (nm). The average wavelength is 550 nm which results in a theoretical limit of resolution (not visibility) of the light microscope in white light of about 200 – 250 nm. The figure below shows two points at the limits of detection and the two individual spots can still be distinguished. The right image shows the two points so close together that the central spots overlap.

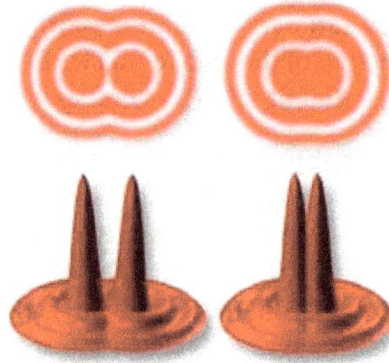

The electron microscope was developed when the wavelength became the limiting factor in light microscopes. Electrons have much shorter wavelengths, enabling better resolution.

## Compare an Optical Microscope vs a Scanning Electron Microscope

As dimensions are shrinking for materials and devices, many structures can no longer be characterized by light microscopy. For example, to determine the integrity of a nanofiber layer for filtration, as shown here, electron microscopy is required to characterize the sample.

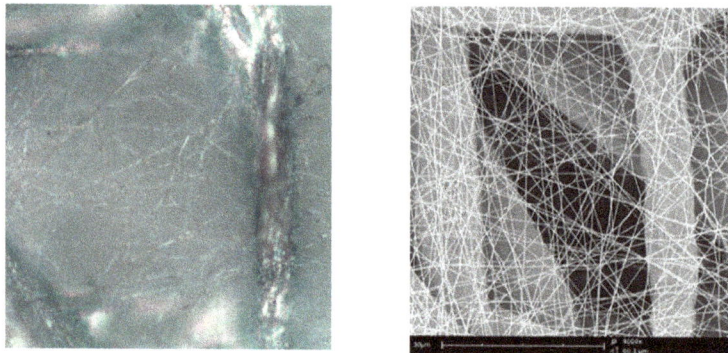

## How a Scanning Electron Microscope Works

The main SEM components include:

- Source of electrons
- Column down which electrons travel with electromagnetic lenses
- Electron detector
- Sample chamber
- Computer and display to view the images

Electrons are produced at the top of the column, accelerated down and passed through a combination of lenses and apertures to produce a focused beam of electrons which hits the surface of the sample. The sample is mounted on a stage in the chamber area and, unless the microscope is designed to operate at low vacuums, both the column and the chamber are evacuated by a combination of pumps. The level of the vacuum will depend on the design of the microscope.

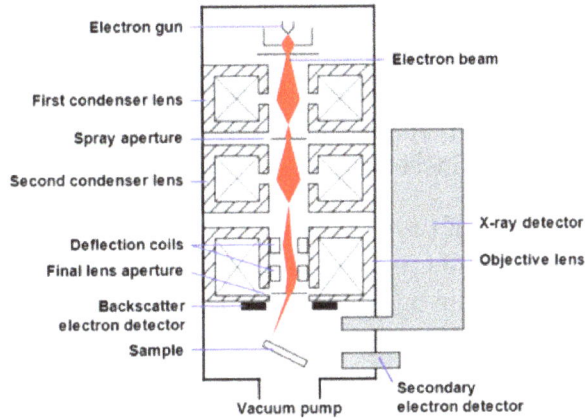

The position of the electron beam on the sample is controlled by scan coils situated above the objective lens. These coils allow the beam to be scanned over the surface of the sample. This beam rastering or scanning, as the name of the microscope suggests, enables information about a defined area on the sample to be collected. As a result of the electron-sample interaction, a number of signals are produced. These signals are then detected by appropriate detectors.

## Sample-electron Interaction

The scanning electron microscope (SEM) produces images by scanning the sample with a high-energy beam of electrons. As the electrons interact with the sample, they produce secondary electrons, backscattered electrons, and characteristic X-rays. These signals are collected by one or more detectors to form images which are then displayed on the computer screen. When the electron beam hits the surface of the sample, it penetrates the sample to a depth of a few microns, depending on the accelerating voltage and the density of the sample. Many signals, like secondary electrons and X-rays, are produced as a result of this interaction inside the sample.

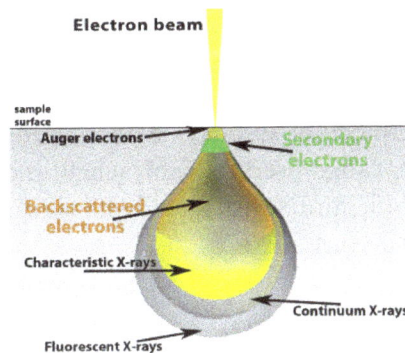

The maximum resolution obtained in an SEM depends on multiple factors, like the electron spot size and interaction volume of the electron beam with the sample. While it cannot provide atomic resolution, some SEMs can achieve resolution below 1 nm. Typically, modern full-sized SEMs provide resolution between 1-20 nm whereas desktop systems can provide a resolution of 20 nm or more.

## Environmental SEM

Conventional SEM requires samples to be imaged under vacuum, because a gas atmosphere rapidly spreads and attenuates electron beams. As a consequence, samples that produce a significant amount of vapour, e.g. wet biological samples or oil-bearing rock, must be either dried or cryogenically frozen. Processes involving phase transitions, such as the drying of adhesives or melting of alloys, liquid transport, chemical reactions, and solid-air-gas systems, in general cannot be observed. Some observations of living insects have been possible however.

The first commercial development of the ESEM in the late 1980s allowed samples to be observed in low-pressure gaseous environments (e.g. 1–50 Torr or 0.1–6.7 kPa) and high relative humidity (up to 100%). This was made possible by the development of a secondary-electron detector capable of operating in the presence of water vapour and by the use of pressure-limiting apertures with differential pumping in the path of the electron beam to separate the vacuum region (around the gun and lenses) from the sample chamber.

The first commercial ESEMs were produced by the ElectroScan Corporation in USA in 1988. ElectroScan was taken over by Philips (who later sold their electron-optics division to FEI Company) in 1996.

ESEM is especially useful for non-metallic and biological materials because coating with carbon or gold is unnecessary. Uncoated Plastics and Elastomers can be routinely examined, as can uncoated biological samples. Coating can be difficult to reverse, may conceal small features on the surface of the sample and may reduce the value of the results obtained. X-ray analysis is difficult with a coating of a heavy metal, so carbon coatings are routinely used in conventional SEMs, but ESEM makes it possible to perform X-ray microanalysis on uncoated non-conductive specimens; however some specific for ESEM artifacts are introduced in X-ray analysis. ESEM may be the preferred for electron microscopy of unique samples from criminal or civil actions, where forensic analysis may need to be repeated by several different experts.

It is possible to study specimens in liquid with ESEM or with other liquid-phase electron microscopy methods.

## Transmission SEM

The SEM can also be used in transmission mode by simply incorporating an appropriate detector below a thin specimen section. Both bright and dark field imaging has been reported in the generally low accelerating beam voltage range used in SEM, which increases the contrast of unstained biological specimens at high magnifications with a field emission electron gun. This mode of operation has been abbreviated by the acronym tSEM.

## Color in SEM

Electron microscopes do not naturally produce color images, as an SEM produces a single value per pixel; this value corresponds to the number of electrons received by the detector during a small period of time of the scanning when the beam is targeted to the (x, y) pixel position.

This single number is usually represented, for each pixel, by a grey level, forming a "black-and-white" image. However, several ways have been used to get color electron microscopy images.

## False Color using a Single Detector

- On compositional images of flat surfaces (typically BSE):

The easiest way to get color is to associate to this single number an arbitrary color, using a color look-up table (i.e. each grey level is replaced by a chosen color). This method is known as false color. On a BSE image, false color may be performed to better distinguish the various phases of the sample.

- On textured-surface images:

As an alternative to simply replacing each grey level by a color, a sample observed by an oblique beam allows researchers to create an approximative topography image. Such topography can then be processed by 3D-rendering algorithms for a more natural rendering of the surface texture.

## SEM Image Coloring

Very often, published SEM images are artificially colored. This may be done for aesthetic effect, to clarify structure or to add a realistic appearance to the sample and generally does not add information about the specimen.

Coloring may be performed manually with photo-editing software, or semi-automatically with dedicated software using feature-detection or object-oriented segmentation.

## Color Built using Multiple Electron Detectors

In some configurations more information is gathered per pixel, often by the use of multiple detectors.

As a common example, secondary electron and backscattered electron detectors are superimposed and a color is assigned to each of the images captured by each detector, with an end result of a combined color image where colors are related to the density of the components. This method is known as density-dependent color SEM (DDC-SEM). Micrographs produced by DDC-SEM retain topographical information, which is better captured by the secondary electrons detector and combine it to the information about density, obtained by the backscattered electron detector.

## Analytical Signals Based on Generated Photons

Measurement of the energy of photons emitted from the specimen is a common method to get analytical capabilities. Examples are the energy-dispersive X-ray spectroscopy (EDS) detectors used in elemental analysis and cathodoluminescence microscope (CL) systems that analyse the intensity and spectrum of electron-induced luminescence in (for example) geological specimens. In SEM systems using these detectors it is common to color code these extra signals and superimpose them in a single color image, so that differences in the distribution of the various components of the specimen can be seen clearly and compared. Optionally, the standard secondary electron image can be merged with the one or more compositional channels, so that the specimen's structure and composition can be compared. Such images can be made while maintaining the full integrity of the original signal data, which is not modified in any way.

## 3D in SEM

SEMs do not naturally provide 3D images contrary to SPMs. However 3D data can be obtained using an SEM with different methods as follows.

### 3D SEM Reconstruction from a Stereo Pair

- photogrammetry is the most metrologically accurate method to bring the third dimension to SEM images. Contrary to photometric methods (next paragraph), photogrammetry calculates absolute heights using triangulation methods. The drawbacks are that it works only if there is a minimum texture, and it requires two images to be acquired from two different angles, which implies the use of a tilt stage. (Photogrammetry is a software operation that calculates the shift (or "disparity") for each pixel, between the left image and the right image of the same pair. Such disparity reflects the local height).

### Photometric 3D SEM Reconstruction from a Four-quadrant Detector "Shape from Shading"

This method typically uses a four-quadrant BSE detector (alternatively for one manufacturer, a 3-segment detector). The microscope produces four images of the same specimen at the same time, so no tilt of the sample is required. The method gives metrological 3D dimensions as far as the slope of the specimen remains reasonable. Most SEM manufacturers now offer such built-in or optional four-quadrant BSE detector, together with proprietary software allowing to calculate a 3D image in real time.

Other approaches use more sophisticated (and sometimes GPU-intensive) methods like the optimal estimation algorithm and offer much better results at the cost of high demands on computing power.

In all instances, this approach works by integration of the slope, so vertical slopes and overhangs are ignored; for instance, if an entire sphere lies on a flat, little more than the upper hemisphere is seen emerging above the flat, resulting in wrong altitude of the sphere apex. The prominence of this effect depends on the angle of the BSE detectors with respect to the sample, but these detectors are usually situated around (and close to) the electron beam, so this effect is very common.

### Photometric 3D Rendering from a Single SEM Image

This method requires an SEM image obtained in oblique low angle lighting. The grey-level is then interpreted as the slope, and the slope integrated to restore the specimen topography. This method is interesting for visual enhancement and the detection of the shape and position of objects ; however the vertical heights cannot usually be calibrated, contrary to other methods such as photogrammetry.

### Other types of 3D SEm Reconstruction

- Inverse reconstruction using electron-material interactive models.
- Vertical stacks of SEM micrographs plus image-processing software.

- Multi-Resolution reconstruction using single 2D File: High-quality 3D imaging may be an ultimate solution for revealing the complexities of any porous media, but acquiring them is costly and time consuming. High-quality 2D SEM images, on the other hand, are widely available. Recently, a novel three-step, multiscale, multiresolution reconstruction method is presented that directly uses 2D images in order to develop 3D models. This method, based on a Shannon Entropy and conditional simulation, can be used for most of the available stationary materials and can build various stochastic 3D models just using a few thin sections.

## Applications of 3D SEM

One possible application is measuring the roughness of ice crystals. This method can combine variable-pressure environmental SEM and the 3D capabilities of the SEM to measure roughness on individual ice crystal facets, convert it into a computer model and run further statistical analysis on the model. Other measurements include fractal dimension, examining fracture surface of metals, characterization of materials, corrosion measurement, and dimensional measurements at the nano scale (step height, volume, angle, flatness, bearing ratio, coplanarity, etc.).

## Coordinate Measuring Machine (CMM)

Co-ordinate Measuring Machines are built rigidly and are very precise. They are equipped with digital readout or can be linked to computers for online inspection of parts. These machines can be placed close to machine tools for efficient inspection and rapid feedback for correction of processing parameter before the next part is made. They are also made more rugged to resist environmental effects in manufacturing plants such as temperature variations, vibration and dirt. Important features of the CMMs are:

(i) To give maximum rigidity to machines without excessive weight, all the moving members, the bridge structure, Z-axis carriage, and Z-column are made of hollow box construction.

(ii) A map of systematic errors in machine is built up and fed into the computer system so that the error compensation is built up into the software.

(iii) All machines are provided with their own computers with interactive dialogue facility and friendly software.

(iv) Thermocouples are incorporated throughout the machine and interfaced with the computer to be used for compensation of temperature gradients and thus provide increased accuracy and repeatability.

A CMM consists of four main elements :

## Main Structure

The machine incorporates the basic concept of three coordinate axes so that precise movement in x, y, and z directions is possible. Each axis is fitted with a linear measurement transducer. The transducers sense the direction of movement and gives digital display. Accordingly, there may be four types of arrangement:

## Cantilever

The cantilever construction combines easy access and relatively small floor space requirements. It is typically limited to small and medium sized machines. Parts larger than the machine table can be inserted into the open side without inhibiting full machine travel. Figure shows a cantilever structure.

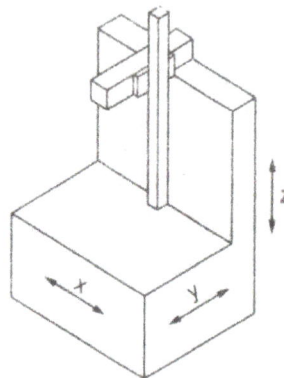

Cantilever Structure

## Bridge Type

The bridge arrangement over the table carries the quill (z-axis) along the x-axis and is sometimes referred to as a travelling bridge. It is claimed that the bridge construction provides better accuracy, although it may be offset by difficulty in making two members track in perfect alignment. This is by far the most popular CMM construction. Figure shows a bridge structure.

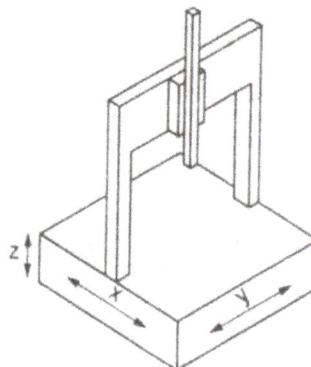

Bridge Structure

## Column Type

The column type machine is commonly referred to as a universal measuring machine rather than a CMM. These machines are usually considered gage room instruments rather than production floor machine. The direction of movements of the arms are as shown in figure. The constructional difference in column type with the cantilever type is with x and y-axes movements.

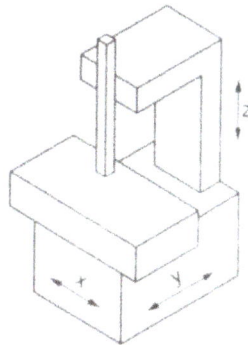

Column Structure

## Gantry

In a gantry type arrangement, arms are held by two fixed supports as shown in figure. Other two arms are capable of sliding over the supports.

Movements of the x, y and z-axes are also as shown in figure below. The gantry type construction is particularly suited for very large components and allows the operator to remain close to the area of inspection.

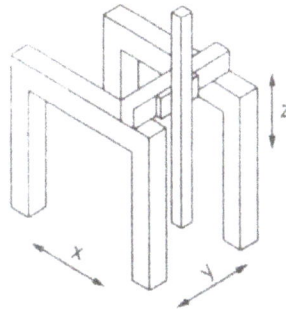

Gantry Structure

## Horizontal

Figure below shows the construction of a horizontal structure. The open structure of this arrangement provides optimum accessibility for large objects such as dies, models, and car bodies. Some horizontal arm machines are referred to as layout machines. There are some horizontal machines where the probe arm can rotate like a spindle to perform tramming operations. Tramming refers to accurate mechanical adjustment of instrument or machine with the help of tram.

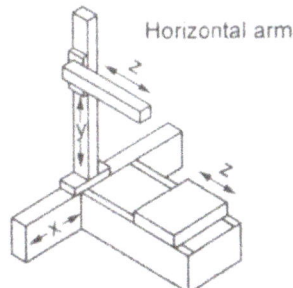

Horizontal Structure

## Probing System

It is the part of a CMM that sense the different parameters required for the calculation. Appropriate probes have to be selected and placed in the spindle of the CMM. Originally, the probes were solid or hard, such as tapered plugs for locating holes. These probes required manual manipulation to establish contact with the workpiece, at which time the digital display was read. Nowadays, transmission trigger-probes, optical transmission probes, multiple or cluster probes, and motorized probes are available. They are discussed in brief below:

## Inductive and Optical Transmission Probes

Inductive and optical transmission probes have been developed for automatic tool changing. Power is transmitted using inductive linking between modules fitted to the machine structure and attached to the probe. figure below shows a schematic of the inductive transmission probe. The hard-wired transmission probe shown is primarily for tool setting and is mounted in a fixed position on the machine structure.

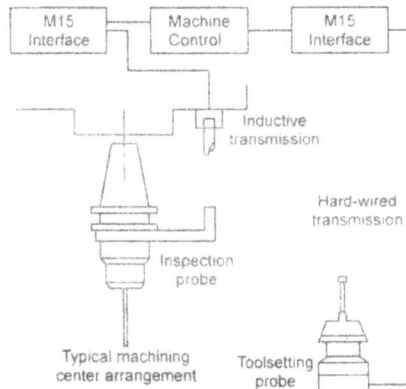

Inductive Probe System and Automatic Probe Changing

The optical transmission probe shown in figure below allows probe rotation between gaging moves, making it particularly useful for datuming the probe. The wide-angle system allows greater axial movement of the probe and is suitable for the majority of installation.

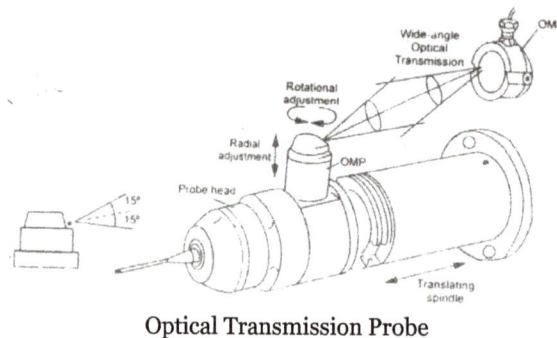

Optical Transmission Probe

## Motorized Probe

With the motorized probe, 48 positions in the horizontal axis, 15 in the vertical axis can be programmed for a total of 720 distinct probe orientations. Figure (b) below shows some typical applications for motorized probe. It shows that with a range of light weight extensions, the head can reach into deep holes and recesses. The second diagram shows that head of the probe is sufficiently compact to be regarded as an extension of the machine quill. This enables the inspection of complex components that would otherwise be impossible or involve complex setups.

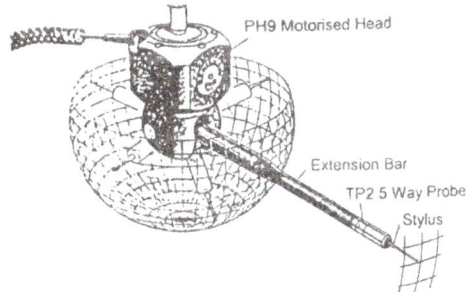

(a) Motorized Probe

(b) Typical Applications of Motorized Probe

## Multiple Styluses Probe Heads

Wide ranges of styli have been developed to suit many different gaging applications. Some of the different styli available are shown mounted on a multiple gaging head in figure below. The selection of stylus is done based on the application for which the probe is to be used.

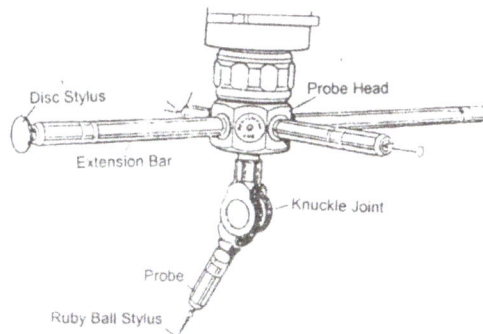

Multiple Stylus Probe Head with Variety of Styli

## Advantages of Cmm

CMM has got a number of advantages. The precision and accuracy given by a CMM is very high. It is because of the inherent characteristics of the measuring techniques used in CMM. Following are the main advantages that CMM can offer:

## Flexibility

CMMs are essentially universal measuring machines and need not be dedicated to any particular task. They can measure almost any dimensional characteristic of a part configuration, including cams, gears and warped surfaces. No special fixtures or gages are required. Because probe contact is light, most parts can be inspected without being clamped to the table.

## Reduced Setup Time

Part alignment and establishing appropriate reference points are very time consuming with conventional surface plate inspection techniques. Software allows the operator to define the orientation of the part on the CMM, and all subsequent data are corrected for misalignment between the parts-reference system and the machine coordinates.

## Single Setup

Most parts can be inspected in a single setup, thus eliminating the need to reorient the parts for access to all features.

## Improved Accuracy

All measurements in a CMM are taken from a common geometrically fixed measuring system, eliminating the introduction and the accumulation of errors that can result with hand-gage inspection methods and transfer techniques.

## Reduced Operator Influence

The use of digital readouts eliminate the subjective interpretation of readings common with dial or vernier type measuring devices. Operator "feel" is virtually eliminated with modern touch-trigger probe systems, and most CMMs have routine measuring procedures for typical part features, such as bores or center distances. In computer assisted systems; the operator is under the control of a program that eliminates operator choice. In addition, automatic data recording, available on most machines, prevents errors in transcribing readings to the inspection report. This adds up to the fact that less skilled operators can be easily instructed to perform relatively complex inspection procedures.

## References

- K. Oura; V. G. Lifshits; A. A. Saranin; A. V. Zotov & M. Katayama (2003). Surface science: an introduction. Berlin: Springer-Verlag. ISBN 978-3-540-00545-2

- Atomic-force-microscope: microscopemaster.com, Retrieved 28 May 2018

- G. Schitter; M. J. Rost (2008). "Scanning probe microscopy at video-rate". Materials Today. 11 (special issue): 40–48. doi:10.1016/S1369-7021(09)70006-9. ISSN 1369-7021. Archived from the original (PDF) on 2009-09-09

- Scanning-tunneling-microscope: britannica.com, Retrieved 17 July 2018

- "Cornell researchers rotate a single molecule of oxygen, making a device that could be used for data storage | Cornell Chronicle". news.cornell.edu. Retrieved 2018-09-07

- Corral, scanning-tunneling-microscopy: nanoscience.com, Retrieved 15 May 2018

- Suzuki, E. (2002). "High-resolution scanning electron microscopy of immunogold-labelled cells by the use of thin plasma coating of osmium". Journal of Microscopy. 208 (3): 153–157. doi:10.1046/j.1365-2818.2002.01082.x

- Scanning-tunneling-microscope-285634: britannica.com, Retrieved 20 April 2018

- R. V. Lapshin (2011). "Feature-oriented scanning probe microscopy". In H. S. Nalwa. Encyclopedia of Nanoscience and Nanotechnology (PDF). 14. USA: American Scientific Publishers. pp. 105–115. ISBN 978-1-58883-163-7

- Scanning-electron-microscopy, techniques: nanoscience.com, Retrieved 30 June 2018

- R. V. Lapshin; O. V. Obyedkov (1993). "Fast-acting piezoactuator and digital feedback loop for scanning tunneling microscopes" (PDF). Review of Scientific Instruments. 64 (10): 2883–2887. Bibcode:1993RScI...64.2883L. doi:10.1063/1.1144377

# 7

# Applications of Nanoparticles

Nanoparticles have potential applications in physics, electronics, optics and medicine. A lot of research is being conducted regarding the use of nanoparticles as potential drug delivery system and as dietary supplements for the delivery of biologically active substances. This is an important chapter, which will analyze in detail about such applications of nanoparticles in diverse industries.

## Applications in Medical Treatment

Medical advancements are constantly being researched. Technologies such as nanoparticles are being used to improve or replace today's therapies. Nanoparticles have advantages over today's therapies because they can be engineered to have certain properties or to behave in a certain way. These properties are selectivity, size, shape, and biocompatibility. Properties such as these allow for nanoparticles to affect the human body differently than traditional therapies. Nanoparticles are being used to increase image contrast of ultrasound and MRI technology. They are used in ultrasound to increase the acoustic reflectivity, ultimately leading to an increase in brightness and creation of a clearer image. Nanoparticles also help increase the image quality of MRI technology. They increase the clarity of the image through the combination of the biocompatible shell and magnetic core. These two components help increase the quality of the image by increasing the half-life of the contrast agent making portions of the MRI targeted by this contrast agent clearer. Nanotechnology helps increase the effectiveness of treating cancer in certain situations. Magnetic field hyperthermia is made more effective by the use of biocompatible super paramagnetic nanoparticles. These particles help heat tissue, particularly cancerous tissue, by creating oscillations that produce heat from friction. This heat will damage the cancerous tissue helping treat the cancer. Cancer is also being treated with gold nanoparticles. These gold nanoparticles are able to select cancerous tissue. Once the tissue has been selected by the particles, a laser is used to explode the particles, which cause damage to the cancerous tissue. The last use of nanoparticles to treat cancer is to increase the effectiveness of radiation therapy; gold nanoparticles are an effective radio sensitizer, are biocompatible and increase dose deposit. Due to the ability of the particles to increase the dose, free radicals are created from the energy absorbed by the particles. These free radicals damage the DNA of cancerous tissues. Not only can nanoparticles be used to treat cancer they can be used to increase drug delivery. Nanoparticles are used in drug delivery because they can be engineered to be sensitive to certain pH values. These particles will remain in a certain conformation protecting the drug till they reach a certain part of the body with a certain pH. In response to the pH the nanoparticles change conformation shape and release the drug. Nanoparticles have also been used to inhibit the bacterial reproduction on surfaces, which creates a cleaner environment and prevents disease.

Lastly, Nanoparticles have been used in orthopedic implants. They increase the biocompatibility of the implants ultimately leading to a longer life span of the implant and effectiveness of the implant. Overall, nanoparticles can be used to create medical advances because of their unique qualities and applications.

## Nanoparticles as Image Enhancing Agents for Ultrasonography

Ultrasonography uses sound waves to create an image for many different purposes. These sound waves are transmitted through the body and bounce off of tissue and return to a receiver. This receiver measures the time it takes for the sound wave to reflect and return to the place of origin, which is perceived as a distance and is converted into an electrical signal, which is then converted into an image by the computer. This type of medical imaging is used in many branches of medicine spanning from obstetrics to oncology. Unfortunately, with ultrasonography minor details could be missed because the image may not be of the best quality. Nanoparticles have been found to help increase the contrast of the image produced by the ultrasonography particularly when imaging tumors. The particles used are called per fluorocarbon emulsion nanoparticles (PFC) and are about 250 nm in diameter. The size of these particles is very important. According to Liu, the smaller particles, particularly ones with surface alteration, have an extended half-life in the circulation and increase the number of passes through tumor vasculature. This is important when helping to get a clear image of a tumor. These particles, due to their size, can be deposited at targeted sites and help increase the acoustic reflectivity because tumor vasculature exhibits an enhanced permeability and retention effect. The increase in reflectivity arises from the difference in acoustic impedance between the tissue and particles. The acoustic reflectivity caused by the retention of the nanoparticles helps to create a better image of the tumor and a better diagnosis. Liu were able to test these hypotheses by performing experiments in which they injected intravenously the nanoparticles in a suspension of saline into mice. By preforming these experiments on a living animal model, their results suggest it could have an application in humans. They were able to conclude from these experiments that the brightness, also called the mean grey scale level, increases with concentration and particle size. The increase in image quality from ultrasonography, due to nanoparticles, can help in many branches of medicine because ultrasonography can be considered more economical than other noninvasive imaging technologies and can help with diagnosis.

## Iron Oxide Nanoparticles for High-Resolution T1 Magnetic Resonance Imaging Contrast Agents

Magnetic Resonance Imaging, otherwise known as MRI, is a widely used technique to non-invasively see into the human body. Even though MRI has improved diagnostic medicine, there can be some problems with it. The images that MRIs produce can sometimes leave out important details. Contrast agents are important in helping to see the details that can sometimes be lost with just MRI technology. "The MRI contrast agents are generally categorized according to their effects on longitudinal (T1) and transversal (T2) relaxations, and their ability is referred to as relaxivity (r1, r2). The area wherein fast T1 relaxation takes place appears bright whereas T2 relaxation results in the dark contrast in the MR images". Increasing the effects of the contrast agents on the T1 relaxation is important because it will create bright spots in the image that could otherwise not have been seen. If these spots cannot be seen a diagnosis could be inconclusive or incorrect. According to Kim, "uniform and extremely small-sized iron oxide nanoparticles" are successful as T1 contrast agents for MRI. The size and shape are important because they affect certain properties of the particle. Larger particles give a larger magnetic effect. The nanoparticles that are used have a biocompatible shell with a magnetic core. The larger the particle the larger the magnetic core within the nanoparticles. These nanoparticles also move and rotate within the fields that they create, which increase T1 effects. In addition to creating a clearer image because of the nanoparticles' size, another reason that they are used as a contrast agent is because these particles also have a long half-life. Iron oxide nanoparticles have been found to have a greater half-life than traditional gadolinium based contrast agents. The optimal size is based upon blood half-life rather than magnetic effect. The nanoparticles "can be good T1 contrast agent for steady-state imaging because they have a long blood half-life derived from their optimal particle size". The longer half-life allows for MRI to be used to track blood pooling in a patient. This is important because it shows that MRI can be used as a more accurate diagnostic tool for problems pertaining to the circulatory system. Overall, iron oxide nanoparticles are successful as a contrast agent because their unique size and half-life allows for improved detail in magnetic resonance images.

## Magnetic Field Hyperthermia: Cancer Treatment with AC Magnetic Field

There are many different treatments for cancer. Hyperthermia is one of the possible treatments being studied. Hyperthermia is the "heating of certain organ or tissues to temperature between 41 °C and 46 °C". By heating tissue within that range it causes damage to the cells because they cannot function at high temperatures. Their lack of function occurs because proteins begin to denature and damage occurs in the cell but not enough to cause cell death. Unfortunately, with hyperthermia, it is difficult to target specific cells, for example cancer cells, without using a targeting agent. To obtain a more concentrated dose of hyperthermia, magnetic particles are used with magnetic fields at specific sites. "Subdomain particles (nanometer in size) absorb much more power at tolerable AC magnetic fields than is obtained by well-known hysteresis heating of multidomain ( microns in size) particles". These nanoparticles have a magnetic core that allows them to develop a magnetic moment. This is important because the magnetic moment of a particle at rest has no specific orientation. Once a magnetic field is applied the particle lines up along the field lines. If the magnetic field is changed the particle will rotate to realign with the new field lines. If the fields are constantly changing as they are with AC magnetic fields then the particle will constantly be rotating from one orientation to another. This oscillation creates a transfer of energy that resembles friction which produces heat. This heat can build up resulting in hyperthermia in the tissue where the magnetic nanoparticles are present. This is a promising treatment for cancer because "a tumor, which has taken up these particles, will not be able to get rid of them". This means if the tumor cells grow between treatments their daughter cells will have some of these nanoparticles inside of them. This implies that only one dose needs to be administered and future applications of AC magnetic fields will affect the daughter cells. Magnetic field hyperthermia is a promising technology because with the magnetic nanoparticles it is more specific to cancer cells and more damaging.

## Laser-induced Explosion of Gold Nanoparticles: Potential Role for Treatment of Cancer

Cancer is a deadly disease, and researchers are constantly looking for a cure or treatment for it. Cancer cells along with bacteria, viruses and DNA can be damaged by nanophotothermolysis with lasers and gold nanoparticles. This is a promising technique because cancer is very invasive and if one cell is left behind it can cause the cancer to regrow. Being able to target individual cells gives a better chance of remission. This technique may also be applied to other diseases in the future because of its ability to target specific cells. "When nanoparticles are irradiated by short laser pulses, their temperature rises very quickly to possibly reach thresholds for nonlinear effects (e.g., microbubble formation, acoustic and shock wave generation) leading to irreparable target (e.g., abnormal cell) damage". The nanoparticle in contact with the cells creates bubbles due to the extreme temperature change. The temperature change is due to the conduction; the particle is excited by the energy conducted from the laser. These bubbles can then burst, sending out shockwaves through the area. The force from the shock wave can disrupt the cell membrane of the target cells damaging them and causing them to lyse. This technique also allows healthy tissue to be spared, because the gold nanospheres can be created to select only cancer cells or other abnormal cells. They select only cancer cells through many different techniques including the use of monoclonal antibodies, recognize the marker found only on the cancer cells and bind to that cell along with the gold nanosphere that it is conjugated to. This specific selection allows for healthy tissue to be spared. Sparing healthy tissue is essential to remission because cancer can damage tissue and cause organ failure. Healthy tissue will at times grow and replace the tissue damaged by cancer. Unfortunately, this technique has low efficiency in "dense solid tumor, bones, atherosclerotic plaques and other targets with a lack of sufficient amount of liquid for efficient bubble generation". The generation of bubbles in the cancer cell from this technique is important because it disrupts the cell, resulting in damage. According to Letfullin, the success of this technique depends on the laser's wavelength, pulse duration, particle size and particle shape. All of these factors are vital in creating localized damage of the cancer cells and sparing healthy tissue.

## Energy Dependence of Gold Nanoparticles Radio Sensitization in Plasmid DNA

Cancer is commonly treated with x-ray radiation. "The aim of radiotherapy is to deliver a lethal dose to tumor volumes while at the same time avoiding exposure to healthy tissue".

Nanoparticles are being used to increase radio sensitization, which is when the cells become more susceptible to radiation damage. "Gold nanoparticles (GNPs) are of considerable interest for use as a radio sensitizer, because of their biocompatibility and their ability to increase dose deposited because of their high mass energy absorption coefficient". Biocompatibility is important because the particles are used within the human body and for them to be beneficial they should not harm healthy tissue. These particles accumulate in tumor cells, which make them useful therapeutic agents for the treatment of cancer. The gold nanoparticles are respectable radiosenitizers because they absorb "10 to 150 time more energy per unit mass than soft tissue". When the particles absorb more energy, more damage is caused to the tumor cells by the ionization of water molecules and creation of free radicals near DNA in the tumor cells. "These radicals cause the majority of DNA damage observed in this system, and as a result, a significant overproduction of these species would lead to a corresponding increase in damage". Radicals are created when gold nanoparticles absorb the energy from radiation, which causes electrons to become excited and to create the free radicals. These free radicals then damage the DNA and cause breakage in the strands. The damage to the DNA inhibits cellular reproduction and growth. Overall, gold nanoparticles are considered promising radiosensitizers because they are selective to tumor cells and cause damage to the DNA within the tumor cells.

## Antimicrobial Effects of Silver Nanoparticles Against Bacterial Cells

Microbes are found everywhere and can be harmful if they find a way to invade the human body. They are becoming even more dangerous because the number of antibiotic resistant microbes is increasing. Nanoparticles are being used to combat the spread of microbes. They are used in the "prevention of bacterial colonization on surfaces of prostheses, catheters, dental materials, and food processing surfaces, such as stainless steel". These are surfaces that are commonly used by humans and could harbor bacteria. Silver nanoparticles have antibacterial properties and are non-toxic to humans in low concentrations. Antibacterial properties inhibit the reproduction of bacteria, which is a microbe. The silver nanoparticles can "inactivate proteins, blocking respiration and electron transfer, and subsequently inactivating the bacteria". Inactivation of the bacteria does not allow them to reproduce and will result in a sanitized surface. The nanoparticles are able to interact with the microbes because the "cell wall peptidoglycans contain negatively charged molecules that will likely interact electrostatically with the silver ions". The silver nanoparticles naturally have a positive charge, which causes them to be attracted to the negatively charged molecules within the cell wall and causes damage to the cell by interruption of its natural processes. The antibacterial properties of the silver nanoparticles depend on the size of the particles; the smaller the particles the greater the effect. The particle size is a major factor because the smaller the particle the greater the surface area, which allows for greater interaction with the bacteria. "Nanoparticles and silver ions interact with sulfur-containing compounds found in bacterial membrane protein and with phosphorous-containing compounds, such as DNA". The interaction with the DNA can also cause a decrease in microbe reproduction, allowing the antimicrobial effects on surfaces to be successful. Inhibiting microbe reproduction with silver nanoparticles decreases the harm that microbes can cause to humans. This can lead to more sterilized environments that could otherwise be overlooked and result in harm.

## Carbon Nanotubes for Orthopaedic Implants

Orthopedic implants are used daily in many surgical procedures. The purpose of an orthopedic implant is to replace or support a damaged bone or joint. These implants are used to help patients achieve a better quality of life. Orthopedic implants are not without limitations. According to Spear and Cameron, "Limitations in structural and biological compatibility with natural bone tissue can cause bone loss from the implantation site and subsequent loosening of the implant, implant failures and complicated revision surgeries." The use of carbon nanotubes in orthopedic implants is being studied as a possible solution for these limitations. Carbon nanotubes, about 0.7-100 nm in length, depending on if they are single wall or multiwall, can be used to improve the lifespan of an orthopedic implant. These nanotubes are made up of carbon atoms and they resemble collagen fibers of regular bone tissue in properties, morphology and dimension. The resemblance of the carbon nanotubes and the collagen fibers allow for improved bone regeneration and adherence to the implant, which in turn creates a more stable implant. Carbon nanotubes have been used in two branches of bone and tissue engineering. They have been used to create mechanical reinforcement for a ceramics and polymers composite used as implants and as coating to improve the surface of the implants so they are more bio-reactive. By helping with mechanical reinforcement, carbon nanotubes can help implants withstand the day to day pressure the human body places on them. In addition, creation of more bio-reactive surfaces can also help with the day to day pressure because the natural bone tissue surrounding the implant would better adhere to it. Both of these uses create a more stable implant overall. Carbon nanotubes in orthopedic implants also have some challenges. One of these challenges is that these coatings and composites may lead to the production of nanotubes as "wear debris", which may trigger an immune response. With any immune response there could be a number of side effects that could limit the effectiveness of these implants. Overall, the use of carbon nanotubes in orthopedic implants is still being researched, and ideas are being tested due to the many positive outcomes that could be occur with their application in bone and tissue engineering.

# Applications in Drug Delivery

In nanotechnology nano particles are used for site specific drug delivery. In this technique the required drug dose is used and side-effects are lowered significantly as the active agent is deposited in the morbid region only. This highly selective approach can reduce costs and pain to the patients. Thus variety of nano particles such as dendrimers, and nano porous materials find application. Micelles obtained from block co-polymers, are used for drug encapsulation. They transport small drug molecules to the desired location. Similarly, nano electromechanical systems are utilized for the active release of drugs. Iron nano particles or gold shells are finding important application in the cancer treatment. A targeted medicine reduces the drug consumption and treatment expenses, making the treatment of patients cost effective.

Nano medicines used for drug delivery, are made up of nano scale particles or molecules which can improve drug bioavailability. For maximizing bioavailability both at specific places in the body and over a period of time, molecular targeting is done by nano engineered devices such as nano robots. The molecules are targeted and delivering of drugs is done with cell precision. In vivo imaging is another area where Nano tools and devises are being developed for in vivo imaging. Using nano particle images such as in ultrasound and MRI, nano particles are used as contrast. The nano engineered materials are being developed for effectively treating illnesses and diseases such as cancer. With the advancement of nanotechnology, self-assembled biocompatible nano devices can be created which will detect the cancerous cells and automatically evaluate the disease, will cure and prepare reports.

The pharmacological and therapeutic properties of drugs can be improved by proper designing of drug delivery systems, by use of lipid and polymer based nano particles. The strength of drug delivery systems is their ability to alter the pharmacokinetics and bio distribution of the drug. Nano particles are designed to avoid the body's defense mechanisms can be used to improve drug delivery. New, complex drug delivery mechanisms are being developed, which can get drugs through cell membranes and into cell cytoplasm, thereby increasing efficiency. Triggered response is one way for drug molecules to be used more efficiently. Drugs that are placed in the body can activate only on receiving a particular signal. A drug with poor solubility will be replaced by a drug delivery system, having improved solubility due to presence of both hydrophilic and hydrophobic environments. Tissue damage by drug can be prevented with drug delivery, by regulated drug release. With drug delivery systems larger clearance of drug from body can be reduced by altering the pharmacokinetics of the drug. Potential nano drugs will work by very specific and well understood mechanisms; one of the major impacts of nanotechnology and Nano science will be in leading development of completely new drugs with more useful behavior and less side effects.

Thus nano particles are promising tools for the advancement of drug delivery, as diagnostic sensors and bio imaging. The bio-distribution of these nanoparticles is still imperfect due to the complex host's reactions to nano- and micro sized materials and the difficulty in targeting specific organs in the body. Efforts are made to optimize and better understand the potential and limitations of nano particulate systems. In the excretory system study of mice dendrimers are encapsulated for drug delivery of positively-charged gold nano particles, which were found to enter the kidneys while negatively-charged gold nanoparticles remained in the important organs like spleen and liver. The positive surface charge of the nanoparticle decreases the rate of opsonization of nanoparticles in the liver, thus affecting the excretory pathway. Due to small size of 5 nm, nano particles can get stored in the peripheral tissues, and therefore can get collected in the body over time. Thus nano particles can be used successfully and efficiently for targeting and distribution, further research can be done on nano toxicity so that its medical uses can be increased and improved.

## The Applications of Nano Particles in Drug Delivery

Abraxane, is albumin bound paclitaxel, a nano particle used for treatment of breast cancer and non-small- cell lung cancer (NSCLC). Nano particles are used to deliver the drug with enhanced effectiveness for treatment for head and neck cancer, in mice model study, which was carried out at from Rice University and University of Texas MD Anderson Cancer Center. The reported treatment uses Cremophor EL which allows the hydrophobic paclitaxel to be delivered intravenously. When the toxic Cremophor is replaced with carbon nano particles its side effects diminished and drug targeting was much improved and needs a lower dose of the toxic paclitaxel.

Nano particle chain was used to deliver the drug doxorubicin to breast cancer cells in a mice study at Case Western Reserve University. The scientists prepared a 100 nm long nano particle chain by chemically linking three magnetic, iron-oxide nano spheres, to one doxorubicinloaded liposome. After penetration of the nano chains inside the tumor magnetic nanoparticles were made to vibrate by generating, radiofrequency field which resulted in the rupture of the liposome, thereby dispersing the drug in its free form throughout the tumor. Tumor growth was halted more effectively by nanotechnology than the standard treatment with doxorubicin and is less harmful to healthy cells as very less doses of doxorubicin were used.

Polyethylene glycol (PEG) nano particles carrying payload of antibiotics at its core were used to target bacterial infection more precisely inside the body, as reported by scientists of MIT. The nano delivery of particles, containing a sub-layer of pH sensitive chains of the amino acid histidine, is used to destroy bacteria that have developed resistance to antibiotics because of the targeted high dose and prolonged release of the drug. Nanotechnology can be efficiently used to treat various infectious diseases.

Researchers in the Harvard University Wyss Institute have used the biomimetic strategy in a mouse model .Drug coated nano particles were used to dissolve blood clots by selectively binding to the narrowed regions in the blood vessels as the platelets do. Biodegradable nano particle aggregates were coated with tissue plasminogen activator, tPA, were injected intravenously which bind and degrade the blood clots. Due to shear stresses in the vessel narrowing region dissociation of the aggregates occurs and releases the tPA-coated nano particles. The nano therapeutics can be applied greatly to reduce the bleeding, commonly found in standard thrombosis treatment.

The researchers in the University of Kentucky have created X-shaped RNA nano particles, which can carry four functional modules. These chemically and thermodynamically stable RNA molecules are able of remaining intact in the mouse body for more than 8 hours and to resist degradation by RNAs in the blood stream. These X-shaped RNA can be effectively performing therapeutic and diagnostic functions. They regulate gene expression and cellular function, and are capable of binding to cancer cells with precision, due to its design.

'Minicell' nano particle are used in early phase clinical trial for drug delivery for treatment of patients with advanced and untreatable cancer. The minicells are built from the membranes of mutant bacteria and were loaded with paclitaxel and coated with cetuximab, antibodies and used for treatment of a variety of cancers. The tumor cells engulf the minicells. Once inside the tumor, the anti-cancer drug destroys the tumor cells. The larger size of minicells plays a better profile in sideeffects. The minicell drug delivery system uses lower dose of drug and has less side-effects can be used to treat a number of different cancers with different anti-cancer drugs.

Nano sponges are important tools in drug delivery, due to their small size and porous nature they can bind poorly-soluble drugs within their matrix and improve their bioavailability. They can be made to carry drugs to specific sites, thus help to prevent drug and protein degradation and can prolong drug release in a controlled manner.

## Proteins and Peptide Delivery

Protein and peptides are macromolecules and are called biopharmaceuticals. These have been identified for treatment of various diseases and disorders as they exert multiple biological actions in human body. Nano materials like nano particles and dendrimers are called as nano biopharmaceuticals , are used for targeted and/or controlled delivery.

## Applications

Nano particles were found useful in delivering the myelin antigens, which induce immune tolerance in a mouse model with relapsing multiple sclerosis. In this technique, biodegradable polystyrene micro particles coated with the myelin sheath peptides will reset the mouse's immune system and thus prevent the recurrence of disease and reduce the symptoms as the protective myelin sheath forms coating on the nerve fibers of the central nervous system. This method of treatment can potentially be used in treatment of various other autoimmune diseases.

## Cancer

Due to the small size of nano particles can be of great use in oncology, particularly in imaging. Nano particles, such as quantum dots, with quantum confinement properties, such as size-tunable light emission, can be used in conjunction with magnetic resonance imaging, to produce exceptional images of tumor sites. As compared to organic dyes, nano particles are much brighter and need one light source for excitation. Thus the use of fluorescent quantum dots could produce a higher contrast image and at a lower cost than organic dyes used as contrast media. But quantum dots are usually made of quite toxic elements.

Nano particles have a special property of high surface area to volume ratio, which allows various functional groups to get attached to a nano particle and thus bind to certain tumor cells. Furthermore, the 10 to 100 nm small size of nanoparticles, allows them to preferentially accumulate at tumor sites as tumors lack an effective lymphatic drainage system. Multifunctional nano particles can be manufactured that would detect, image, and then treat a tumor in future cancer treatment. Kanzius RF therapy attaches microscopic nano particles to cancer cells and then "cooks" tumors inside the body with radio waves that heat only the nanoparticles and the adjacent (cancerous) cells.

Nano wires are used to prepare sensor test chips, which can detect proteins and other biomarkers left behind by cancer cells, and detect and make diagnosis of cancer possible in the early stages from a single drops of a patient's blood.

Nano technology based drug delivery is based upon three facts: i) efficient encapsulation of the drugs, ii) successful delivery of said drugs to the targeted region of the body, and iii) successful release of that drug there.

Nano shells of 120 nm diameter, coated with gold were used to kill cancer tumors in mice by Prof. Jennifer at Rice University. These nano shells are targeted to bond to cancerous cells by conjugating antibodies or peptides to the nano shell surface. Area of the tumor is irradiated with an infrared laser, which heats the gold sufficiently and kills the cancer cells.

Cadmium selenide nano particles in the form of quantum dots are used in detection of cancer tumors because when exposed to ultraviolet light, they glow. The surgeon injects these quantum dots into cancer tumors and can see the glowing tumor, thus the tumor can easily be removed.

Nano particles are used in cancer photodynamic therapy, wherein the particle is inserted within the tumor in the body and is illuminated with photo light from the outside. The particle absorbs light and if it is of metal, it will get heated due to energy from the light. High energy oxygen molecules are produced due to light which chemically react with and destroy tumors cell, without reacting with other body cells. Photodynamic therapy has gained importance as a noninvasive technique for dealing with tumors.

The applications of various nano systems in cancer therapy are summarized as:

- Carbon nano tubes, 0.5–3 nm in diameter and 20–1000 nm length, are used for detection of DNA mutation and for detection of disease protein biomarker.

- Dendrimers, less than 10 nm in size are useful for controlled release drug delivery, and as image contrast agents.

- Nano crystals, of 2-9.5 nm size cause improved formulation for poorly-soluble drugs, labeling of breast cancer marker HeR2 surface of cancer cells.

- Nano particles are of 10-1000 nm size and are used in MRI and ultrasound image contrast agents and for targeted drug delivery, as permeation enhancers and as reporters of apoptosis, angiogenesis.

- Nano shells find application in tumor-specific imaging, deep tissue thermal ablation.

- Nano wires are useful for disease protein biomarker detection, DNA mutation detection and for gene expression detection.

- Quantum dots, 2-9.5 nm in size, can help in optical detection of genes and proteins in animal models and cell assays, tumor and lymph node visualization.

# Nanotechnology in the Treatment of Neurodegenerative Disorders

One of the most important applications of nanotechnology is in the treatment of neuro degenerative disorders. For the delivery of CNS therapeutics, various nano carriers such as, dendrimers, nano gels, nano emulsions, liposomes, polymeric nano particles, solid lipid nano particles, and nano suspensions have been studied. Transportation of these nano medicines has been effected across various in vitro and in vivo BBB models by endocytosis and/or transcytosis, and early preclinical success for the management of CNS conditions such as, Alzheimer's disease, brain tumors, HIV encephalopathy and acute ischemic stroke has become possible. The nanomedicine can be advanced further by improving their BBB permeability and reducing their neurotoxicity.

Parkinson's disease: This can improve current therapy of Parkinson's disease (PD). Parkinson's disease (PD) is the second most common neurodegenerative disease after Alzheimer's disease and affects one in every 100 persons above the age of 65 years, PD is a disease of the central nervous system; neuro inflammatory responses are involved and leads to severe difficulties with body motions. The present day therapies aim to improve the functional capacity of the patient for as long as possible but cannot modify the progression of the neurodegenerative process.

Aim of applied nanotechnology is regeneration and neuro protection of the central nervous system (CNS) and will significantly benefit from basic nanotechnology research conducted in parallel with advances in neurophysiology, neuropathology and cell biology. The efforts are taken to develop novel technologies that directly or indirectly help in providing neuro protection and/or a permissive environment and active signaling cues for guided axon growth. In order to minimize the peripheral side-effects of conventional forms of Parkinson's disease therapy, research is focused on the design, biometric simulation and optimization of an intracranial nano-enabled scaffold device (NESD) for the site-specific delivery of dopamine to the brain, as a strategy. Peptides and peptidic nano particles are newer tools for various CNS diseases.

Nanotechnology will play a key role in developing new diagnostic and therapeutic tools. Nanotechnology could provide devices to limit and reverse neuro pathological disease states, to support and promote functional regeneration of damaged neurons, to provide neuro protection and to facilitate the delivery of drugs and small molecules across the blood–brain barrier. For the delivery of CNS therapeutics, various nano carriers such as dendrimers, nano gels, nano emulsions, liposomes, polymeric nano particles, solid lipid nano particles, and nano suspensions have been studied. Transportation of these nano medicines has been effected across various in vitro and in vivo BBB models by endocytosis and/or transcytosis, and early preclinical success for the management of CNS conditions such as, Alzheimer's disease, brain tumors, HIV encephalopathy and acute ischemic stroke has become possible. Future development of CNS Nano medicines needs to focus on increasing their drug-trafficking performance and specificity for brain tissue using novel targeting moieties.

Alzheimer's disease: Worldwide, more than 35 million people are affected by Alzheimer's disease (AD), which is the most common form dementia. Nano technology finds significant applications in neurology. These approaches are based on the, early AD diagnosis and treatment is made possible by designing and engineering of a plethora of nanoparticulate entities with high specificity for brain capillary endothelial cells. Nano particles (NPs) have high affinity for the circulating amyloid-β (Aβ) forms and therefore may induce "sink effect" and improve the AD condition. In vitro diagnostics for AD has advanced due to ultrasensitive NP-based bio-barcodes and immune sensors, as well as scanning tunneling microscopy procedures capable of detecting Aβ1–40 and Aβ1–42. The recent research on use of nano particles in the treatment of Alzheimer's disease.

Tuberculosis treatment: Tuberculosis (TB) is the deadly infectious disease. The long duration of the treatment and the pill burden can hamper patient lifestyle and result in the development of multi-drug resistant (MDR) strains. Tuberculosis in children constitutes a major problem. There is commercial non availability of the first-line drugs in pediatric form. Novel antibiotics can be designed to overcome drug resistance, cut short the duration of the treatment course and to reduce drug interactions with antiretroviral therapies. A nanotechnology is one of the most promising approaches for the development of more effective and compliant medicines. The advancements in nanobased drug delivery systems for encapsulation and release of anti-TB drugs can lead to development of a more effective and affordable TB pharmacotherapy.

## The Clinical Application of Nanotechnology in Operative Dentistry

Nanotechnology aims at the creation and utilization of materials and devices at the atomic, and molecular level, supra molecular structures, and in the exploitation of unique properties of particles of size 0.1 nm to 100 nm. Nano filled composite resin materials are believed to offer excellent wear resistance, strength, and ultimate aesthetics due to their exceptional polishability and luster retention. In operative dentistry, nano fillers constitute spherical silicon dioxide ($SiO_2$) particles with an average size of 5-40 nm. The real innovation about nano fillers is the possibility of improving the load of inorganic phase. The effect of this high filler load is widely recorded in terms of mechanical properties. Micro hybrid composites with additional load of Nano fillers are the best choice in operative dentistry. It is expected that in near future, it would be possible to use a filler material in operative dentistry, whose shape and composition would closely mimic the optical and mechanical characteristics of the natural hard tissues (enamel and dentin). It also explains the basic concepts of fillers in composite resins, scanning electron microscopy and energy dispersive spectroscopy evaluation, and filler weight content. Nano composite resins are non-agglomerated discrete nanoparticles that are homogeneously distributed in resins or coatings to produce nano composites have been successfully manufactured by nano products Corporation. The nanofiller used is aluminosilicate powder with a mean particle size of 80 ran 1:4 M ratio of alumina to silica and a refractive index of 1.508. These nano composites have superior hardness, flexural strength, modulus of elasticity, decreased polymerization shrinkage and also have excellent handling properties particle size of 80 ran 1:4 M ratio of alumina to silica and a refractive index of 1.508.

## Applications in Ophthalmology

The aim of nano medicine is the to monitor, control, construct, repair, defense, and improve human biological systems at the molecular level, with the help of nano devices and nanostructures that operate massively in parallel at the unit cell level, in order to achieve medical benefit. Principles of nanotechnology are applied to nano medicine such as bio mimicry and pseudo intelligence. Some applications of nanotechnology to ophthalmology are include treatment of oxidative stress; measurement of intraocular pressure; the agnostics; use of nano particles for treatment of choroidal new vessels, to prevent scars after glaucoma surgery, and for treatment of retinal degenerative disease using gene therapy; prosthetics; and regenerative nano medicine. The current therapeutic challenges in drug delivery, postoperative scarring will be revolutionized with the help of nanotechnology and will help in various unsolved problems such as sight-restoring therapy for patients with retinal degenerative disease. Treatments for ophthalmic diseases are expected from this emerging field. A novel nanoscale dispersed eye ointment (NDEO) for the treatment of severe evaporative dry eye has been successfully developed. The excipients used as semisolid lipids were petrolatum and lanolin, as used in conventional eye ointment, which were coupled with medium-chain triglycerides (MCT) as a liquid lipid; both phases were then dispersed in polyvinyl pyrrolidone solution to form nano dispersion. A transmission electron micrograph showed that the ointment matrix was entrapped in the nano emulsion of MCT, with a mean particle size of about 100 nm. The optimized formulation of NDEO was stable when stored for six months at 4 °C, and demonstrated no cytotoxicity to human corneal epithelial cells when compared with commercial polymer-based artificial tears. The therapeutic effects of NDEO were evaluated and demonstrated therapeutic improvement, displaying a trend of positive correlation with higher concentrations of ointment matrix in the NDEO formulations compared to a marketed product. Histological evaluation demonstrated that the NDEO restored the normal corneal and conjunctival morphology and is safe for ophthalmic application. Recent research shows applications of various nano particulate systems like microemulsions, nanosuspensions, nanoparticles, liposomes, niosomes, dendrimers and cyclodextrins in the field of ocular drug delivery and also depicts how the various upcoming of nanotechnology like nanodiagnostics, nano imaging and nano medicine can be utilized to explore the frontiers of ocular drug delivery and therapy.

## Surgery

The technique developed by Rice University, two pieces of chicken meat is fused by a flesh welder, by placing two pieces of chicken touching each other. In this technique, green liquid containing goldcoated nano shells is allowed to dribble along the seam and two sides are weld together. This method can be used arterieswhich have been cut during organ transplant. The flesh welder can be used to weld the artery perfectly.

## Visualization

Drug distribution and its metabolismcan be determined by tracking movement. Cells were dyed by scientists to track their movement throughout the body. These dyes excited by light of a certain wavelength to glow. Luminescent tags were used to dye various numbers of cells. These tags are quantum dots attached to proteins which penetrate cell membranes. The dots were of various sizes and bio-inert material. As a result, sizes are selected so that the frequency of light used to make a group of quantum dots fluoresce, and used to make another group incandesce. Thus both groups can be lit with a single light source.

## Tissue Engineering

In tissue engineering, nanotechnology can be applied to reproduce or repair damaged tissues. By using suitable nanomaterial-based scaffolds and growth factors, artificially stimulated cell proliferation, in organ transplants or artificial implants therapy nano technology can be useful, which can lead to life extension.

## Antibiotic Resistance

Antibiotic resistance can be decreased by use of nano particles in combination therapy. Zinc Oxide nano particles can decrease the antibiotic resistance and enhance the antibacterial activity of Ciprofloxacin against microorganism, by interfering with various proteins that are interacting in the antibiotic resistance or pharmacologic mechanisms of drugs.

## Immune Response

The nano device bucky balls have been used to alter the allergy/ immune response. They prevent mast cells from releasing histamine into the blood and tissues, as these bind to free radicals better than any anti-oxidant available, such as vitamin E.

## Nano Pharmaceuticals

Nano pharmaceuticals can be used to detect diseases at much earlier stages and the diagnostic applications could build upon conventional procedures using nanoparticles. Nano pharmaceuticals are an emerging field where the sizes of the drug particle or a therapeutic delivery system work at the nanoscale. Delivering the appropriate dose of a particular active agent to specific disease site still remains difficult in the pharmaceutical industry. Nano pharmaceuticals have enormous potential in addressing this failure of traditional therapeutics which offers site-specific targeting of active agents. Nano pharmaceuticals can reduce toxic systemic side effects thereby resulting in better patient compliance.

Pharmaceutical industry faces enormous pressure to deliver highquality products to patients while maintaining profitability. Therefore pharmaceutical companies are using nanotechnology to enhance the drug formulation and drug target discovery. Nano pharmaceutical makes the drug discovery process cost effective, resulting in the improved Research and Development success rate, thereby reducing the time for both drug discovery and diagnostics.

# Applications in Mechanical Engineering

## Spark Plug

Since Nano materials are stronger, harder, and resist wear and erosion, they are currently being considered for use in spark plugs. Nano electrodes would make spark plugs long lasting and fuel-efficient. The rail plug made by Nano material creates powerful sparks that burn fuel better.

## Engine Coatings

Automobiles waste huge amounts of energy by way of heat loss from the engine, especially from the diesel ones. Engineers are currently looking at coating engine cylinders with Nano crystalline ceramics, like zirconia and alumina, to help preserve heat efficiency and increase fuel combustion. Engineers are currently looking at coating engine cylinders with Nano crystalline ceramics, like zirconia and alumina, to help preserve heat efficiency and increase fuel combustion.

## Engine

Today's car engines are only 25 per cent efficient; meaning only a quarter of the energy stored in fuel is actually converted to useful work. Fuel cells - devices that work by harnessing the chemical attraction between oxygen and hydrogen to produce electricity - are, by contrast, 50 per cent efficient. Further, because they use oxygen - which is taken from the air, and hydrogen - the most abundant element in the universe, they have the potential to produce clean and cheap energy. The only by-products are heat and water.

## Braking System, Body Panel

Use of aluminium nanotube composite in the braking system results effective braking performance. It will reduce brake system weight while increasing acceleration. In addition to being lighter, Nano composites are significantly more resistant to wear and tear in case of body panels.

## Chassis

Chassis are structural support of a vehicle. If we can reduce its weight it will become more fuel efficient. So we can use Nano size steel instead of Aluminium. It will offer lower weight, dimensional accuracy, corrosion resistance and aesthetics.

## Application in the Field Of Energy

### Batteries

With the growth in portable electronic equipment (mobile phones, laptop computers), there is great demand for lightweight, high- energy density batteries. Nickel–metal hydride batteries made of Nano crystalline nickel and metal hydrides are requiring less frequent recharging and to last longer because of their large surface area. Nanoparticle matrices in battery electrodes can drastically increase their ability to store lithium ions, increasing the storage density of the battery.

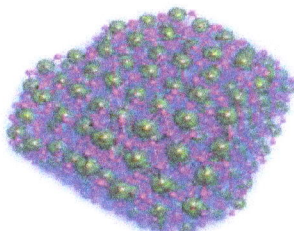

Lithium ions in a battery

The high-rate discharge capability of the Nano materials has also been demonstrated in prototype cells for the Li4Ti5O12 materials. Emerging applications have steered Lithium-ion materials R&D in a new direction, which includes development of nanomaterial electrodes. , Lithium- ion Nano materials can also be expected to appear in automotive applications like PHEV and also in battery electrical energy storage systems.

## Solar Panels

Nanotechnology could help increase the efficiency of light conversion by using nanostructures. Conventional semiconductors manage to use 40% of the Sun's energy. Nano materials help in reducing materials and process rates, energy saving and enhanced renewable energy sources.

## Application in the Field of Construction

Steel has been widely available material and has a major role in the construction industry. The theoretical strength of steel is 27.30 GPa (in <111> direction). There are two ways of achieving high strength in steels. One method is by reducing the size of a crystal to such an extent that it is free of any defects. Second method is by introducing a very large density of defects in a metal sample that act as an obstacle to the motion of dislocations.

The Nano size steel produce stronger steel cables which can be used in bridge construction. The strengthening arises due to the presence of Nano scale cementite/ferrite lamellar structure. The high carbon steel wire is an important engineering material used for reinforcing automobile tires, galvanized wires etc.

## Steel

Steel is a widely available material that has a major role in the construction industry. The use of nanotechnology in steel helps to improve the physical properties of steel. Fatigue, or the structural failure of steel, is due to cyclic loading. Current steel designs are based on the reduction in the allowable stress, service life or regular inspection regime. This has a significant impact on the life-cycle costs of structures and limits the effective use of resources. Advancements in this technology through the use of nanoparticles would lead to increased safety, less need for regular inspection, and more efficient materials free from fatigue issues for construction.

Steel cables can be strengthened using carbon nanotubes. Stronger cables reduce the costs and period of construction, especially in suspension bridges, as the cables are run from end to end of the span.

The use of vanadium and molybdenum nanoparticles improves the delayed fracture problems associated with high strength bolts. This reduces the effects of hydrogen embrittlement and improves steel micro-structure by reducing the effects of the inter-granular cementite phase.

## Glass

Research is being carried out on the application of nanotechnology to glass, another important material in construction. Titanium dioxide (TiO2) nanoparticles are used to coat glazing since it has sterilizing and anti-fouling properties. The particles catalyse powerful reactions that break down organic pollutants, volatile organic compounds and bacterial membranes. TiO2 is hydrophilic (attraction to water), which can attract rain drops that then wash off the dirt particles. Thus the introduction of nanotechnology in the Glass industry.

# Applications in Manufacturing and Materials

Many benefits of nanotechnology depend on the fact that it is possible to tailor the structures of materials at extremely small scales to achieve specific properties, thus greatly extending the materials science toolkit. Using nanotechnology, materials can effectively be made stronger, lighter, more durable, more reactive, more sieve-like, or better electrical conductors, among many other traits. Many everyday commercial products are currently on the market and in daily use that rely on nanoscale materials and processes:

- Nanoscale additives to or surface treatments of fabrics can provide lightweight ballistic energy deflection in personal body armor, or can help them resist wrinkling, staining, and bacterial growth.

- Clear nanoscale films on eyeglasses, computer and camera displays, windows, and other surfaces can make them water- and residue-repellent, antireflective, self-cleaning, resistant to ultraviolet or infrared light, antifog, antimicrobial, scratch-resistant, or electrically conductive.

- Nanoscale materials are beginning to enable washable, durable "smart fabrics" equipped with flexible nanoscale sensors and electronics with capabilities for health monitoring, solar energy capture, and energy harvesting through movement.

- Lightweighting of cars, trucks, airplanes, boats, and space craft could lead to significant fuel savings. Nanoscale additives in polymer composite materials are being used in baseball bats, tennis rackets, bicycles, motorcycle helmets, automobile parts, luggage, and power tool housings, making them lightweight, stiff, durable, and resilient. Carbon nanotube sheets are now being produced for use in next-generation air vehicles. For example, the combination of light weight and conductivity makes them ideal for applications such as electromagnetic shielding and thermal management.

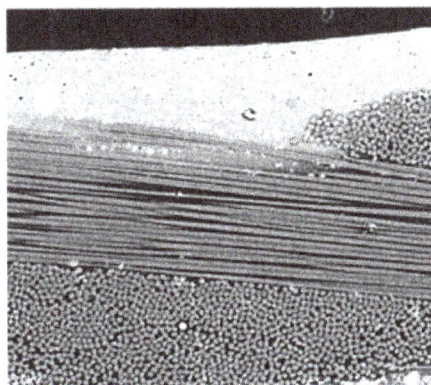

High-resolution image of a polymer-silicate nanocomposite. This material has improved thermal, mechanical, and barrier properties and can be used in food and beverage containers, fuel storage tanks for aircraft and automobiles, and in aerospace components.

- Nano-bioengineering of enzymes is aiming to enable conversion of cellulose from wood chips, corn stalks, unfertilized perennial grasses, etc., into ethanol for fuel. Cellulosic nano-materials have demonstrated potential applications in a wide array of industrial sectors, including electronics, construction, packaging, food, energy, health care, automotive, and defense. Cellulosic nanomaterials are projected to be less expensive than many other nano-materials and, among other characteristics, tout an impressive strength-to-weight ratio.

- Nano-engineered materials in automotive products include high-power rechargeable battery systems; thermoelectric materials for temperature control; tires with lower rolling resistance; high-efficiency/low-cost sensors and electronics; thin-film smart solar panels; and fuel additives for cleaner exhaust and extended range.

- Nanostructured ceramic coatings exhibit much greater toughness than conventional wear-resistant coatings for machine parts. Nanotechnology-enabled lubricants and engine oils also significantly reduce wear and tear, which can significantly extend the lifetimes of moving parts in everything from power tools to industrial machinery.

- Nanoparticles are used increasingly in catalysis to boost chemical reactions. This reduces the quantity of catalytic materials necessary to produce desired results, saving money and reducing pollutants. Two big applications are in petroleum refining and in automotive catalytic converters.

- Nano-engineered materials make superior household products such as degreasers and stain removers; environmental sensors, air purifiers, and filters; antibacterial cleansers; and specialized paints and sealing products, such a self-cleaning house paints that resist dirt and marks.

- Nanoscale materials are also being incorporated into a variety of personal care products to improve performance. Nanoscale titanium dioxide and zinc oxide have been used for years in sunscreen to provide protection from the sun while appearing invisible on the skin.

# Applications in the Environment

Nanotechnology is being used in several applications to improve the environment. This includes cleaning up existing pollution, improving manufacturing methods to reduce the generation of new pollution, and making alternative energy sources more cost effective.

In trying to help our ailing environment, nanotechnology researchers and developers are pursuing the following avenues:

Generating less pollution during the manufacture of materials. One example of this is how researchers have demonstrated that the use of silver nano clusters as catalysts can significantly reduce the polluting byproducts generated in the process used to manufacture propylene oxide. Propylene oxide is used to produce common materials such as plastics, paint, detergents and brake fluid.

Producing solar cells that generate electricity at a competitive cost. Researcher has demonstrated that an array of silicon nanowires embedded in a polymer results in low cost but high efficiency solar cells. This, or other efforts using nanotechnology to improve solar cells, may result in solar cells that generate electricity as cost effectively as coal or oil.

Increasing the electricity generated by windmills. Epoxy containing carbon nanotubes is being used to make windmill blades. The resulting blades are stronger and lower weight and therefore the amount of electricity generated by each windmill is greater.

Cleaning up organic chemicals polluting groundwater. Researchers have shown that iron nanoparticles can be effective in cleaning up organic solvents that are polluting groundwater. The iron nanoparticles disperse throughout the body of water and decompose the organic solvent in place. This method can be more effective and cost significantly less than treatment methods that require the water to be pumped out of the ground.

Cleaning up oil spills. Using photo catalytic copper tungsten oxide nanoparticles to break down oil into biodegradable compounds. The nanoparticles are in a grid that provides high surface area for the reaction, is activated by sunlight and can work in water, making them useful for cleaning up oil spills.

Clearing volatile organic compounds (VOCs) from air. Researchers have demonstrated a catalyst that breaks down VOCs at room temperature. The catalyst is composed of porous manganese oxide in which gold nanoparticles have been embedded.

Reducing the cost of fuel cells. Changing the spacing of platinum atoms used in a fuel cell increases the catalytic ability of the platinum. This allows the fuel cell to function with about 80% less platinum, significantly reducing the cost of the fuel cell.

Storing hydrogen for fuel cell powered cars. Using graphene layers to increase the binding energy of hydrogen to the graphene surface in a fuel tank results in a higher amount of hydrogen storage and a lighter weight fuel tank. This could help in the development of practical hydrogen-fueled cars.

## Applications in Electronics

Nanotechnology has greatly contributed to major advances in computing and electronics, leading to faster, smaller, and more portable systems that can manage and store larger and larger amounts of information. These continuously evolving applications include:

- Transistors, the basic switches that enable all modern computing, have gotten smaller and smaller through nanotechnology. nanometer transistor in 2016! Smaller, faster, and better transistors may mean that soon your computer's entire memory may be stored on a single tiny chip.

- Using magnetic random access memory (MRAM), computers will be able to "boot" almost instantly. MRAM is enabled by nanometer-scale magnetic tunnel junctions and can quickly and effectively save data during a system shutdown or enable resume-play features.

- Ultra-high definition displays and televisions are now being sold that use quantum dots to produce more vibrant colors while being more energy efficient.

SUNY College of Nanoscale Science and Engineering's Michael Liehr, left, and
IBM's Bala Haranand display a wafer comprised of 7nm chips in a NFX clean room in Albany.

- Flexible, bendable, foldable, rollable, and stretchable electronics are reaching into various sectors and are being integrated into a variety of products, including wearables, medical applications, aerospace applications, and the Internet of Things. Flexible electronics have been developed using, for example, semiconductor nanomembranes for applications in smartphone and e-reader displays. Other nanomaterials like graphene and cellulosic nano-materials are being used for various types of flexible electronics to enable wearable and "tattoo" sensors, photovoltaics that can be sewn onto clothing, and electronic paper that can be rolled up. Making flat, flexible, lightweight, non-brittle, highly efficient electronics opens the door to countless smart products.

- Other computing and electronic products include Flash memory chips for smart phones and thumb drives; ultra-responsive hearing aids; antimicrobial/antibacterial coatings on keyboards and cell phone casings; conductive inks for printed electronics for RFID/smart cards/smart packaging; and flexible displays for e-book readers.

- Nanoparticle copper suspensions have been developed as a safer, cheaper, and more reli-able alternative to lead-based solder and other hazardous materials commonly used to fuse electronics in the assembly process.

## Applications in Solar Energy Harvesting

Solar energy harvesting and storage have nowadays attracted tremendous research efforts due to the increasing energy challenges that fossil fuels are being rapidly consumed by modern technolo-gy. For material chemists, how to develop materials that are effective in harnessing solar energy is the primary step to tackle this challenge. Apart from selection of materials with intrinsic physical properties to utilize solar energy, scientists have also been studying the unique properties when the particle size reduced to the nanometer scale. Nanomaterials, which have high surface-to-vol-ume ratio, cover a variety of shapes of nanoparticles, nanorods, nanoporous framework, and so on. One can also easily tune the optical and charge transfer properties by changing the size of semicon-ductor nanomaterials. The chemical properties such as catalytic activity can also be remarkably changed with increased surface atoms of nanocatalysts.

This special issue is mainly dedicated to the synthesis of functional nanomaterials which can be used to harvest or store solar energy.

The desire to make solar energy into usable energy format has led to the fast development of solar cells, a direct way to convert solar energy into electricity. The ideal candidate materials to be used in solar cells should possess proper band gap to harvest as much sunlight as possible, good charge transport properties, excellent stability, and cheap cost. R.-F. Guan and coworkers prepared the Cu-$InS_2$ thin films via one-step electrodeposition method. They investigated the effects of synthesis parameters on the structure, morphology, and optical and electrical properties on the final $CuInS_2$ thin films, and optimal preparation conditions were proposed. , R.-F. Guan and coworkers used magnetron sputtering followed by sulfurization method to prepare $CuInS_2$ thin films. Optimized sputtering power, sputtering gas pressure, heat treatment temperature, and duration were reported.

As a special type of solar cell, Dye-Sensitized Solar Cells (DSSCs) use photoexcited electrons generated from the conduction band of $TiO_2$ thin films to generate electricity with the assistance of dye molecules. The key factors to prepare an effective DSSC are high surface area of $TiO_2$ thin film, efficient charge injection from dye molecules to $TiO_2$, and minimal loss of electrons from $TiO_2$ to outer circuit. Kim and coworkers developed $TiO_2$ nanobranch/nanoparticle hybrid structure to increase the surface area of thin film and thus the loading of dye molecules. Light harvesting efficiency is greatly promoted, leading to tripled incident photon conversion efficiency (IPCE) compared with bare nanobranch thin films.

Another alternative and interesting approach to utilize solar energy is through the processes of photocatalysis or artificial photosynthesis to convert solar energy into chemical energy to drive certain chemical reactions, for example, water splitting or photodegradation of organic pollutants. P. Dong and coworkers synthesized Ag3PO$_4$ triangular prism by a simple coprecipitation method which was used as an excellent photocatalyst for organic compounds degradation under light irradiation.

Biomass, the nature's way of storing solar energy, is also covered in our special issue. Y. Zhang and coworkers used corn straw as the raw material, ionic liquids as solvents, and acids as catalysts to study the effects of hydrolysis conditions on reducing sugar concentration, and optimal hydrolysis parameters were provided.

White light emitting diodes (LEDs) have been acknowledged as highly efficient and green light sources. When coupled with solar cells, the white LEDs can provide efficient lighting without consuming other energy resources. R.-F. Guan and coworkers synthesized $Y_{2.94-x}Al_5O_{12}(YAG):Ce_{0.06}Pr_x$ phosphors with various $Pr^{3+}$ concentrations by coprecipitation method. They investigated the influence of $Pr^{3+}$ doping concentrations on the phases, luminescent properties, and energy transfer phenomenon from $Ce^{3+}$ to $Pr^{3+}$.

# Permissions

# Index